André Schleife

Electronic and optical properties of MgO, ZnO, and CdO

André Schleife

Electronic and optical properties of MgO, ZnO, and CdO

Parameter-free calculation of real-structure effects in three transparent conductive oxides

Südwestdeutscher Verlag für Hochschulschriften

Impressum/Imprint (nur für Deutschland/only for Germany)
Bibliografische Information der Deutschen Nationalbibliothek: Die Deutsche Nationalbibliothek verzeichnet diese Publikation in der Deutschen Nationalbibliografie; detaillierte bibliografische Daten sind im Internet über http://dnb.d-nb.de abrufbar.
Alle in diesem Buch genannten Marken und Produktnamen unterliegen warenzeichen-, marken- oder patentrechtlichem Schutz bzw. sind Warenzeichen oder eingetragene Warenzeichen der jeweiligen Inhaber. Die Wiedergabe von Marken, Produktnamen, Gebrauchsnamen, Handelsnamen, Warenbezeichnungen u.s.w. in diesem Werk berechtigt auch ohne besondere Kennzeichnung nicht zu der Annahme, dass solche Namen im Sinne der Warenzeichen- und Markenschutzgesetzgebung als frei zu betrachten wären und daher von jedermann benutzt werden dürften.

Coverbild: www.ingimage.com

Verlag: Südwestdeutscher Verlag für Hochschulschriften GmbH & Co. KG
Dudweiler Landstr. 99, 66123 Saarbrücken, Deutschland
Telefon +49 681 37 20 271-1, Telefax +49 681 37 20 271-0
Email: info@svh-verlag.de

Zugl.: Jena, FSU, Diss., 2010

Herstellung in Deutschland:
Schaltungsdienst Lange o.H.G., Berlin
Books on Demand GmbH, Norderstedt
Reha GmbH, Saarbrücken
Amazon Distribution GmbH, Leipzig
ISBN: 978-3-8381-2766-8

Imprint (only for USA, GB)
Bibliographic information published by the Deutsche Nationalbibliothek: The Deutsche Nationalbibliothek lists this publication in the Deutsche Nationalbibliografie; detailed bibliographic data are available in the Internet at http://dnb.d-nb.de.
Any brand names and product names mentioned in this book are subject to trademark, brand or patent protection and are trademarks or registered trademarks of their respective holders. The use of brand names, product names, common names, trade names, product descriptions etc. even without a particular marking in this works is in no way to be construed to mean that such names may be regarded as unrestricted in respect of trademark and brand protection legislation and could thus be used by anyone.

Cover image: www.ingimage.com

Publisher: Südwestdeutscher Verlag für Hochschulschriften GmbH & Co. KG
Dudweiler Landstr. 99, 66123 Saarbrücken, Germany
Phone +49 681 37 20 271-1, Fax +49 681 37 20 271-0
Email: info@svh-verlag.de

Printed in the U.S.A.
Printed in the U.K. by (see last page)
ISBN: 978-3-8381-2766-8

Copyright © 2011 by the author and Südwestdeutscher Verlag für Hochschulschriften GmbH & Co. KG and licensors
All rights reserved. Saarbrücken 2011

Für meine Familie,
Für Oma Ruth.
Für Yvonne.

We've stuck to our own beliefs,
we haven't cheated anyone,
and we've done what we wanted.

Lars Ulrich

Contents

1 Introduction — 7

2 Fundamentals — 11
  2.1 Setting the stage — 11
    2.1.1 Matter — 11
    2.1.2 Interacting electrons — 12
    2.1.3 Quantum-field theoretical description — 13
  2.2 Ground state: Density functional theory — 14
    2.2.1 Hohenberg-Kohn theorem I — 15
    2.2.2 Hohenberg-Kohn theorem II — 16
    2.2.3 Kohn-Sham equations — 16
    2.2.4 Exchange and correlation — 19
    2.2.5 Non-collinear spins — 21
  2.3 One-particle excitations — 23
    2.3.1 Green's function and equation of motion — 23
    2.3.2 The electronic self-energy — 25
  2.4 Two-particle excitations — 26
    2.4.1 Bethe-Salpeter equation — 27
    2.4.2 Excitonic Hamiltonian — 28
    2.4.3 Macroscopic dielectric function — 30
    2.4.4 Screening in heavily doped materials — 30
    2.4.5 Semiconductor Bloch equations — 32
  2.5 Alloy statistics and thermodynamics — 34
    2.5.1 Cluster expansion — 34
    2.5.2 Generalized quasi-chemical approximation — 35
    2.5.3 Strict-regular solution and microscopic decomposition limit — 37

3 Practical issues — 39
  3.1 Electronic properties — 40
    3.1.1 Hybrid functional and quasiparticle corrections — 40
    3.1.2 Mapping to an affordable approach — 41
    3.1.3 Inclusion of spin-orbit coupling — 41
  3.2 Optical properties — 42
    3.2.1 Adapted sampling of the Brillouin zone — 42
    3.2.2 Inclusion of spin-orbit coupling — 43
    3.2.3 Screening of the electron-hole interaction — 44

4 Ideal MgO, ZnO, and CdO — 45
4.1 One-particle excitations — 46
4.1.1 Band structures and densities of states — 46
4.1.2 Inclusion of spin-orbit coupling — 52
4.1.3 Application: Band alignment at interfaces — 56
4.2 Two-particle excitations — 59
4.2.1 Impact of many-body effects on the optical properties — 59
4.2.2 Complex frequency-dependent dielectric function — 61
4.2.3 Excitons and spin-orbit coupling — 66
4.2.4 Application: Electron-energy loss function — 68
4.3 Summary — 70

5 Lattice distortions: Strain and non-equilibrium polymorphs — 71
5.1 Uniaxial and biaxial strain in ZnO — 72
5.1.1 Quasiparticle energies in the proximity of the band gap — 73
5.1.2 Excitons under the influence of biaxial strain — 74
5.2 Non-equilibrium wurtzite structure: MgO and CdO — 75
5.2.1 Quasiparticle energies — 76
5.2.2 Optical properties of the absorption edge — 78
5.3 Summary — 80

6 Pseudobinary alloys: Isostructural versus heterostructural MgZnO and CdZnO — 81
6.1 Thermodynamic properties and lattice structure — 82
6.1.1 Mixing free energy — 82
6.1.2 Structural composition of heterostructural alloys — 86
6.2 One-particle excitations — 86
6.2.1 Quasiparticle band structures — 87
6.2.2 Densities of states — 90
6.3 Dielectric function of wz-$Mg_xZn_{1-x}O$ — 93
6.4 Summary — 94

7 A point defect: The oxygen vacancy as F-center in rs-MgO — 95
7.1 Atomic geometries and charge states — 96
7.2 Transition energies and absorption — 97
7.3 Exciton binding energies — 98
7.4 Summary — 99

8 Heavy n-doping: Wannier-Mott and Mahan excitons in wz-ZnO — 101
8.1 Approaching the problem via a two-band model — 102
8.1.1 Effects due to a degenerate electron gas — 102
8.1.2 Semiconductor Bloch equations — 103
8.2 Ab-$initio$ calculations for wz-ZnO — 106
8.2.1 Absorption coefficient — 107
8.2.2 Binding energies and oscillator strengths — 108
8.2.3 Inter-conduction-band absorption — 110

	8.3	Summary	111
9	**The end ... and future prospects**	**113**
A	**Appendix**	**115**
	A.1	Cluster expansions for the wurtzite and the rocksalt crystal structure	115
	A.2	Parameters used in the calculations	115

Bibliography   **121**

1 Introduction

> Und aus dem Chaos sprach eine Stimme zu mir: "Lächle und sei froh, es könnte schlimmer kommen!" Und ich lachte und war froh – denn es kam schlimmer!
>
> Otto Waalkes

Thousands of years ago, in the glowing embers of the dawning Bronze age, *alloying* the two metals copper and tin was a ground-breaking discovery that started a new era. Centuries later, gaining an understanding of matter had become a central goal of the *philosophy of nature* and the application of this knowledge has acted as basis of the *progress* for mankind since then. Historically, physics was an empirical field, continuously accompanied by efforts to achieve reliable *predictions*. In the beginning of the last century, with the advent of quantum theory, the fundament for an atomistic description of matter was laid. It was clear from the very beginning that the corresponding equations are too complex to be solved exactly for real systems. Approximations had to be made, and, ironically, are the reason why theory and experiment started from different points of view. Available samples of the materials were far from the ideal systems that theory was able to describe. Nowadays, generations later, both disciplines have approached each other. Enhanced experimental techniques provide crystals of high quality, while the theoretical description benefits from the continuously increasing power of computers, which renders them capable of solving complicated problems without crude approximations.

Interestingly, computers are not just providing solutions for existing problems. Their increasing capabilities triggered the evolution from the industrial towards the information age and they even became an own driving force for development, e.g., materials research. Initially, the electronic circuits that allowed the breakthrough of the computer were largely silicon-based. Nowadays, the next wave of this development is about to roll down — mobility. Mobile devices working with fast wireless networks enable the Internet to become an integrated part of our lives. However, they have slightly different requirements than traditional computers. A truly independent power supply raises demands for new energy sources, such as solar cells. We even hope to harness the movement of the human body to generate electricity someday, e.g. via piezoelectric ZnO nanowires [1]. In addition, such integration into everyday life imposes demands on the user interface of such devices. Brilliant displays, as in windshields or glasses, require transparent electronics. When silicon reaches its limits, new materials pitch in.

The motivation for the present work emerged from both of the aforementioned developments; we employ recent *ab-initio* methods and theoretical spectroscopy techniques that rely on heavy numerical calculations to describe electronic excitations in non-ideal crystals of three group-II oxides. While zinc oxide (ZnO) is already widely applied for optoelectronics (see e.g. Ref. [2]), magnesium- (MgO), and cadmium oxide (CdO) are possible candidates for combinations with ZnO, for instance in alloys or heterostructures. We study the ideal equilibrium polymorphs of these oxides, that are experimentally well-investigated, for gaining a thorough understanding as well as the necessary confidence in our approaches to generalize and apply them to the electronic excitations in imperfect crystals. As such *imperfections* we take the influence of strain, the alloying of the different oxides, an intrinsic point defect, and free electrons in the lowest conduction band into account. Strain that occurs for instance when thin films are grown on a lattice-mismatched substrate is experimentally relevant due to its impact on the electronic and optical properties. Alloys of the three oxides are of highest interest in the context of band-gap tailoring. In the case of rs-MgO the oxygen vacancy, as a prototypical F-center, raises questions concerning the respective optical absorption peaks for decades. Finally, in the context of transparent and conductive materials the influence of the free carriers on the optical properties and the bound excitonic states has never been explained by parameter-free calculations. In this work we introduce the respective generalizations of the theoretical and numerical approaches and perform the computationally involved calculations.

Along the path from quantum mechanics towards actual calculations, density functional theory (DFT) [3, 4] is a milestone since it provides access to the ground-state properties of materials. Probing the electronic band structure experimentally, e.g. by means of spectroscopy techniques, corresponds to adding an electron or a hole to the system. Taking the response of the electronic system to this excitation into account in the calculations, leads to the quasiparticle picture that can be described using Hedin's GW approximation [5, 6] for the electronic self energy. We employ the DFT results as input in order to compute quasiparticle electronic structures, which are in good agreement with experimental findings. Moreover, optical measurements typically create electron-hole pairs in the system. According to Hedin's equations for interacting electrons [5, 6], the electron-hole interaction is taken into account by solving a Bethe-Salpeter equation for the polarization function. This quantity is related to the dielectric function, which allows us to access the optical properties of the oxides. In

Chapter 2 we introduce these theoretical concepts as well as the generalization of the Bethe-Salpeter approach to account for a partially occupied conduction band. Remarks concerning the practical calculations are discussed in Chapter 3.

Chapter 4 describes the equilibrium polymorphs of ideal bulk MgO, ZnO, and CdO and investigates the structure of their valence and conduction bands. We present densities of states and effective masses, as well as natural band discontinuities. Furthermore, our description of the dielectric function, which takes excitonic effects into account, enables us to derive the electron-energy loss function. Throughout, detailed comparisons to experimental results prove the suitability of our parameter-free theoretical approaches.

Evidently, *ab-initio* calculations can provide insight beyond experimentally accessible parameter ranges. In this context, the influence of uniaxial and biaxial strain on the ordering of the valence bands in ZnO is investigated in Chapter 5. In addition, we explore the electronic band structure of the non-equilibrium wurtzite structures of MgO and CdO, for which no bulk crystals exist, preventing an experimental investigation. Hence, we *predict* valence-band splittings and band gaps as they might occur at interfaces of MgO or CdO with ZnO substrates.

In Chapter 6 we study pseudobinary alloys by means of a cluster expansion method. Appropriate cluster statistics allow us to elaborate on the impact of different growth conditions on the composition of the alloy. Due to the different crystal structures of the respective oxides, i.e. rocksalt and wurtzite, the description of their *heterostructural* combination has to be achieved. The electronic and optical properties of the group-II oxide alloys are calculated and discussed with respect to different growth conditions. The corresponding calculations are computationally extremely expensive due to the large number of possible clusters.

The oxygen vacancy in MgO is studied in Chapter 7. Along these lines, the resemblance between the absorption peaks of the F-center and the F^+-center is most puzzling. We will show how the inclusion of excitonic effects in the many-body calculations allows us to unravel experimental observations even though the solution of the Bethe-Salpeter equation for supercells containing a defect is extremely demanding from a computational point of view. Besides, the investigation of the F^+-center requires a spin-polarized treatment of the excitonic Hamiltonian which only recently became possible.

Turning a transparent material conductive by introducing free electrons via heavy *n*-doping is essential e.g. for photovoltaic applications. In Chapter 8 we calculate the frequency-dependent absorption of ZnO, accounting for the first time for excitonic effects and free electrons in the lowest conduction band within a first-principles framework. The Bethe-Salpeter approach has to be extended to account for the partially occupied conduction-band states and for the impact on the screening of the electron-hole interaction. Thereby, we disentangle the interplay of both aspects and explain how they affect the optical-absorption properties. Furthermore, we explore the possibility of an excitonic Mott transition. These investigations are computationally expensive since they require the calculation of highly accurate exciton binding energies.

Finally, we summarize our insights regarding the influence of imperfections on the group-II oxides in Chapter 9. We contribute to explanations of experimental findings, leading to a deeper understanding of these oxides. Aside from possibly emerging applications, we must not forget the spirit of solving problems within collaborations which was a major driving force behind the present work.

2 Fundamentals

> Nothing exists except atoms and empty space; everything else is opinion.
>
> Democritus

2.1 Setting the stage

2.1.1 Matter

For thousands of years mankind has been challenged to understand what forms their environment and the world around them. Many generations of scientists and philosophers struggled to contribute piece by piece to what our present grasp of *matter* is. Nowadays, our conception of the building blocks of all that surrounds us relies on an atomistic picture. Every material, be it rigid, liquid, or gaseous, is built of atoms. Likewise, the atoms themselves show a substructure, consisting of a heavy, positively charged nucleus, and a certain number of negative electrons around it. This complex of the core and its electrons, bound together via the Coulomb force, forms the electrostatically neutral atom. Looking deeper into this structure and, therefore, going well above the energy scale of the Coulomb interaction, scientists found that the nucleus itself consists again of different, even more elementary particles. It seems to be a fundamental principle that, at least to some extent, our view of the world strongly depends on the energy scale we are using to look at it.

Condensed-matter physics is the superordinate framework of this work. Its experimental techniques typically do not interfere with the nuclear structure of matter since the characteristic energies are too low, being in the range of less than a milli-electron volt (meV) up to several 10 keV. The dominant interaction is the Coulomb force which leads to the negative electrons and the positive cores attracting each other. In contrast, this force causes repulsion between the electrons and between the nuclei. While this sounds like a fairly complete picture, the situation is incredibly complicated for a macroscopic system with typically on the order of 10^{23} atoms per cm^3 whose electrons and nuclei potentially all interact. Moreover, this problem has to be treated on a quantum-mechanical footing, i.e., exchange and correlation (XC) enter, going beyond the classic repulsion of the electrons. Fortunately, we are still able to *predict* properties of such systems from first principles, however, sophisticated approximations seem inevitable.

2.1.2 Interacting electrons

In this work we aim to describe the *electronic structure* and the *optical properties* of the three group-II oxides MgO, ZnO, and CdO. Therefore, as a first major simplification, we employ the Born-Oppenheimer approximation [7], according to which we keep only the electronic problem from the total Hamiltonian. Treating the nuclei merely as a static external Coulomb potential for the interacting electrons leads to a neglect of any dynamic interaction between the cores and electrons. The impact of this drastic approximation on the electronic and optical properties will be pointed out where necessary.

For the remaining *electronic problem* we rely on developments dating back to the first third of the 20th century. During this time our picture of the electron dramatically changed when quantum mechanics and the theory of relativity together culminated in the Dirac equation [8, 9]. This equation is, strictly speaking, the solid theoretical ground for the concept of *electron spin*. Since the three oxides of interest in this work have even numbers of electrons, it is reasonable to consider spin-paired electrons only — leading to an entirely spin-less description by means of the Schrödinger equation which is solved for doubly occupied states. For a more complete picture, we are also occasionally interested in the electronic fine structure. In these cases the Pauli equation [10], as the *weakly relativistic* limit of the Dirac equation, will be employed. In addition, the spin-orbit coupling (SOC) term that results from the Dirac description will be included to deal with the interaction of the spin, as an internal angular momentum, with the orbital angular momentum (cf. Section 2.2.5).

For materials with a non-ferromagnetic ground state it is well-justified to neglect the transversal interaction of the electrons, i.e. the vector potential, in the Hamiltonian of the electronic problem [11], which leaves three terms that are taken into account: (i) the electronic kinetic energy $T(\mathbf{r})$, (ii) the external potential $V(\mathbf{r},\mathbf{R})$ caused by the positively charged nuclei, and (iii) the electron-electron interaction $U(\mathbf{r})$. This leads to the Hamilton operator

$$\hat{H}(\mathbf{r},\mathbf{R}) = \hat{T}(\mathbf{r}) + \hat{U}(\mathbf{r}) + \hat{V}(\mathbf{r},\mathbf{R}). \tag{2.1}$$

When describing a system of N electrons (mass m, $\mathbf{r} = \{\mathbf{r}_1,\ldots,\mathbf{r}_N\}$) and M cores (mass M_s,

2.1 Setting the stage

$\mathbf{R} = \{\mathbf{R}_1,\ldots,\mathbf{R}_M\}$, charge Z_s, $s = 1,\ldots,M$) within first quantization these terms are [11]

$$\hat{T}(\mathbf{r}) = \sum_{i=1}^{N} \frac{\hat{\mathbf{p}}_i^2}{2m}, \qquad (2.2)$$

$$\hat{U}(\mathbf{r}) = \frac{1}{2} \cdot \frac{e^2}{4\pi\varepsilon_0} \sum_{\substack{i,j=1 \\ i\neq j}}^{N} \frac{1}{|\hat{\mathbf{r}}_i - \hat{\mathbf{r}}_j|}, \qquad (2.3)$$

$$\hat{V}(\mathbf{r},\mathbf{R}) = -\frac{e^2}{4\pi\varepsilon_0} \sum_{i=1}^{N}\sum_{s=1}^{M} \frac{Z_s}{|\hat{\mathbf{r}}_i - \mathbf{R}_s|}. \qquad (2.4)$$

2.1.3 Quantum-field theoretical description

In a quantum-mechanical description the electrons of the system are indistinguishable. To incorporate this fundamental property into the problem, it is required that the square of the many-electron wave function remains unchanged under any operator that only exchanges two electrons. More specifically, the wave function for fermions (therefore also for electrons) has to be antisymmetric under such a transformation [11]. This property of the wave function $|\psi_\alpha^R\rangle$ (in Dirac's Braket notation) will be ensured by anti-commutator relations for the field operators $\hat{\psi}_\alpha^\dagger(\mathbf{r})$ and $\hat{\psi}_\alpha(\mathbf{r})$. The *creation operator* $\hat{\psi}_\alpha^\dagger(\mathbf{r})$ is defined as the operator that transforms an N particle state into an $(N+1)$ particle state by adding an electron with spin α at the position $\mathbf{r}$. Its adjoint, the *annihilation operator* $\hat{\psi}_\alpha(\mathbf{r}) = \left(\hat{\psi}_\alpha^\dagger(\mathbf{r})\right)^\dagger$, transforms an N electron state to an $(N-1)$ electron state by removing a particle. Constructing the wave function by successive application of creators and annihilators automatically guarantees antisymmetry when these anti-commutator relations are fulfilled [11]:

$$\left[\hat{\psi}_\alpha(\mathbf{r}), \hat{\psi}_\beta(\mathbf{r}')\right]_+ = \left[\hat{\psi}_\alpha^\dagger(\mathbf{r}), \hat{\psi}_\beta^\dagger(\mathbf{r}')\right]_+ = 0 \text{ and} \qquad (2.5)$$

$$\left[\hat{\psi}_\alpha(\mathbf{r}), \hat{\psi}_\beta^\dagger(\mathbf{r}')\right]_+ = \delta(\mathbf{r}-\mathbf{r}')\delta_{\alpha\beta}. \qquad (2.6)$$

Any one- and two-particle operator can be expanded in terms of the field operators via

$$\sum_{i=1}^{N} \hat{A}_1^i = \sum_{\alpha,\beta} \iint d\mathbf{r}_1 d\mathbf{r}_2 \langle \mathbf{r}_1\alpha|\hat{A}_1|\mathbf{r}_2\beta\rangle \hat{\psi}_\alpha^\dagger(\mathbf{r}_1)\hat{\psi}_\beta(\mathbf{r}_2) \text{ and} \qquad (2.7)$$

$$\frac{1}{2}\sum_{i\neq j} \hat{A}_2^{i,j} = \frac{1}{2}\sum_{\alpha,\beta,\gamma,\delta} \iiiint d\mathbf{r}_1 d\mathbf{r}_2 d\mathbf{r}_3 d\mathbf{r}_4 \times \qquad (2.8)$$
$$\times \langle \mathbf{r}_1\alpha, \mathbf{r}_2\beta|\hat{A}_2|\mathbf{r}_3\gamma, \mathbf{r}_4\delta\rangle \hat{\psi}_\alpha^\dagger(\mathbf{r}_1)\hat{\psi}_\beta^\dagger(\mathbf{r}_2)\hat{\psi}_\delta(\mathbf{r}_4)\hat{\psi}_\gamma(\mathbf{r}_3).$$

Applying these transformations to the operators $\hat{T}$, $\hat{U}$, and $\hat{V}$ [Eqs. (2.2), (2.3), (2.4)] yields

$$\hat{T} = -\frac{\hbar^2}{2m} \sum_\alpha \int d\mathbf{r}\, \hat{\psi}_\alpha^\dagger(\mathbf{r})\, \Delta\, \hat{\psi}_\alpha(\mathbf{r}), \tag{2.9}$$

$$\hat{U} = \frac{1}{2} \sum_{\alpha,\beta} \iint d\mathbf{r}_1\, d\mathbf{r}_2\, u(\mathbf{r}_1,\mathbf{r}_2)\, \hat{\psi}_\alpha^\dagger(\mathbf{r}_1)\, \hat{\psi}_\beta^\dagger(\mathbf{r}_2)\, \hat{\psi}_\beta(\mathbf{r}_2)\, \hat{\psi}_\alpha(\mathbf{r}_1), \text{ and} \tag{2.10}$$

$$\hat{V} = \sum_\alpha \int d\mathbf{r}\, v(\mathbf{r},\mathbf{R})\, \hat{\psi}_\alpha^\dagger(\mathbf{r})\, \hat{\psi}_\alpha(\mathbf{r}), \tag{2.11}$$

where the matrix elements $u(\mathbf{r}_1,\mathbf{r}_2)$ and $v(\mathbf{r},\mathbf{R})$ are given by

$$u(\mathbf{r}_1,\mathbf{r}_2) = u(|\mathbf{r}_1 - \mathbf{r}_2|) = \frac{e^2}{4\pi\varepsilon_0 |\mathbf{r}_1 - \mathbf{r}_2|} \text{ and} \tag{2.12}$$

$$v(\mathbf{r},\mathbf{R}) = -\frac{e^2}{4\pi\varepsilon_0} \sum_{s=1}^{M} \frac{Z_s}{|\mathbf{r} - \mathbf{R}_s|} = -\sum_{s=1}^{M} Z_s u(\mathbf{r},\mathbf{R}_s). \tag{2.13}$$

2.2 Ground state: Density functional theory

In Eqs. (2.1)–(2.4) the Hamiltonian of the interacting many-electron problem was introduced, though its solution yet has to be found. The incredibly large number of involved electrons renders an exact treatment of the problem impossible. As pointed out by W. Kohn [12], the reason is not only the dimension of the respective parameter space which even grows exponentially with the particle number N of the problem, but also the unmanageably large amount of information contained in the fully interacting many-body wave function. In 1964 P. Hohenberg and W. Kohn found an appealing and instructive formulation of the problem [3] based on the density $n(\mathbf{r})$ of the ground state $|\psi\rangle$ of the electronic system,

$$n(\mathbf{r}) = \sum_\alpha \langle \psi | \hat{\psi}_\alpha^\dagger(\mathbf{r}) \hat{\psi}_\alpha(\mathbf{r}) | \psi \rangle. \tag{2.14}$$

By proving that $n(\mathbf{r})$, a simple function of three spatial coordinates, can replace the complicated many-body wave function (which depends on three spatial coordinates for each electron) as the basis variable of the problem, they achieved a tremendous conceptual simplification. Furthermore, they provided an approach to calculate the ground-state energy E of a given Hamiltonian $\hat{H}[n]$ via a variational principle. These concepts shall now be further elucidated.

2.2.1 Hohenberg-Kohn theorem I

In the following we consider the Hamiltonian given by Eq. (2.1). We want to remark that in the original paper, Ref. [3], the operator $\hat{V}$ represents *any* one-particle external potential of the type (2.11) for the electronic system and is only in a special case caused by the positive nuclei.

The first Hohenberg-Kohn (HK) theorem establishes a one-to-one mapping between the external potential $v(\mathbf{r},\mathbf{R})$ and the ground-state density of the Hamiltonian. One direction of the proof is simple: Obviously, the density, being the ground-state expectation value of the density operator [cf. Eq. (2.14)], is a functional of the external potential $v(\mathbf{r},\mathbf{R})$. For the reverse direction one has to show that $v(\mathbf{r},\mathbf{R})$ is a unique functional of the ground-state density. Here, we follow the proof by P. Hohenberg and W. Kohn [3] for non-degenerate ground states.

Therefore, $\hat{V}$ and $\hat{V}'$ represent two external potentials that differ by more than just a constant, i.e., $\hat{V} - \hat{V}' \neq \text{const.}$ The corresponding Hamiltonians $\hat{H}$ and $\hat{H}'$ lead, via the Schrödinger equations, to different ground states $|\psi\rangle$ and $|\psi'\rangle$ with the ground-state energies E and E'. For the indirect proof we assume that both potentials may lead to equal ground-state densities n. The variational principle of J. Rayleigh and W. Ritz gives for both Hamiltonians

$$E = \langle\psi|\hat{H}|\psi\rangle < \langle\psi'|\hat{H}|\psi'\rangle = \langle\psi'|\hat{H}' - \hat{V}' + \hat{V}|\psi'\rangle = E' + \langle\psi'|\hat{V} - \hat{V}'|\psi'\rangle, \text{ and} \quad (2.15)$$

$$E' = \langle\psi'|\hat{H}'|\psi'\rangle < \langle\psi|\hat{H}'|\psi\rangle = \langle\psi|\hat{H} - \hat{V} + \hat{V}'|\psi\rangle = E + \langle\psi|\hat{V}' - \hat{V}|\psi\rangle. \quad (2.16)$$

In the sum of Eqs. (2.15) and (2.16) the two terms $\langle\psi'|\hat{V} - \hat{V}'|\psi'\rangle$ and $\langle\psi|\hat{V}' - \hat{V}|\psi\rangle$ cancel each other, as they are expectation values of the type (2.11), due to the assumed equality of the ground-state densities. This then leads to the contradiction

$$E' + E < E + E',$$

which proves the assumption of equality of the ground-state densities to be wrong.

In summary, the first HK theorem states that a given ground-state density $n(\mathbf{r})$ uniquely defines the external potential $v(\mathbf{r},\mathbf{R})$ (except for an irrelevant additive constant), and, therefore, the entire Hamiltonian. By means of the Schrödinger equation this basis-variable property of $n(\mathbf{r})$ transfers to the entire many-body ground state $|\psi\rangle$, i.e.,

$$|\psi\rangle = |\psi[n]\rangle, \; \hat{H} = \hat{H}[n], \; E = \langle\psi[n]|\hat{H}[n]|\psi[n]\rangle = E[n]. \quad (2.17)$$

These relations impressively show that all quantities which can be derived from the Hamiltonian are implicitly contained in its ground-state density [13]. So far, this is only a formal simplification since neither the functionals for calculating any property of a system, nor the recipe for how to construct them, are known *a priori*. We will discuss such an approach for how to obtain the ground-state energy, though it remains questionable if such functionals can be constructed in explicit terms for every excited state energy [13]. In practice, DFT is preferably applied to the calculation of ground-state properties, such as total energies and lattice geometries, and as such is the method we employ in this work.

2.2.2 Hohenberg-Kohn theorem II

For all possible ground states $|\psi[\tilde{n}]\rangle$ the energy $E[\tilde{n}]$ of a system with the ground-state density n is given as the expectation value of its Hamiltonian $\hat{H}[n]$. Due to the first HK theorem $E[\tilde{n}]$ is a unique functional of $\tilde{n}$,

$$E[\tilde{n}] = \langle \psi[\tilde{n}] | \hat{H}[n] | \psi[\tilde{n}] \rangle. \tag{2.18}$$

The second HK theorem states that the basis-variable property of $n(\mathbf{r})$ can also be transferred to a variational principle, i.e., the functional (2.18) assumes its minimum at the ground-state density n of $\hat{H}[n]$. Using the Rayleigh-Ritz variational principle we know that the ground-state wave function $|\psi[n]\rangle$ of the respective system minimizes the energy. For every other ground-state wave function $|\psi[\tilde{n}]\rangle$ it holds

$$\langle \psi[n] | \hat{H}[n] | \psi[n] \rangle = E[\psi[n]] = E[n] < \langle \psi[\tilde{n}] | \hat{H}[n] | \psi[\tilde{n}] \rangle = E[\psi[\tilde{n}]] = E[\tilde{n}]. \tag{2.19}$$

Using the first part of the HK theorem, which ensures the unique mapping between the wave function and the ground-state density, we have now shown that the second HK theorem transforms the variation with respect to $|\psi\rangle$ into a variation with respect to $\tilde{n}(\mathbf{r})$. Extensions of the proofs given here, especially for degenerate ground states, are available in the literature [13].

2.2.3 Kohn-Sham equations

So far it has been shown that due to the HK theorem the ground-state density of the interacting many-electron problem acts as the basis variable of a variational principle for the ground-state energy. For the actual *determination* of this ground-state density, we follow the approach of W. Kohn and L. J. Sham [4] by mapping the problem of interacting electrons in an external potential $v(\mathbf{r}, \mathbf{R})$ onto a system of non-interacting particles (index s) in an *effective*

2.2 Ground state: Density functional theory

potential $v_{\text{eff}}(\mathbf{r},\mathbf{R})$ with the same ground-state density and the energy functional

$$E_s[\tilde{n}(\mathbf{r})] = \int v_{\text{eff}}(\mathbf{r},\mathbf{R})\,\tilde{n}(\mathbf{r})\,d\mathbf{r} + T_s[\tilde{n}(\mathbf{r})]. \tag{2.20}$$

Here, $T_s[\tilde{n}(\mathbf{r})]$ is the universal functional for the kinetic energy of non-interacting particles. The minimization of Eq. (2.20) with respect to the density yields

$$0 = \frac{\delta}{\delta \tilde{n}(\mathbf{r})}\left[E_s[\tilde{n}(\mathbf{r})] - \zeta \int \tilde{n}(\mathbf{r})\,d\mathbf{r}\right]_{\tilde{n}(\mathbf{r})=n_s(\mathbf{r})} = \left[v_{\text{eff}}(\mathbf{r},\mathbf{R}) + \frac{\delta T_s[\tilde{n}(\mathbf{r})]}{\delta \tilde{n}(\mathbf{r})}\right]_{\tilde{n}(\mathbf{r})=n_s(\mathbf{r})} - \zeta \tag{2.21}$$

where the Lagrange multiplier ζ ensures the conservation of the number of particles. From this equation the definition of the potential $v_{\text{eff}}(\mathbf{r},\mathbf{R})$ follows as

$$[v_{\text{eff}}(\mathbf{r},\mathbf{R})]_{\tilde{n}(\mathbf{r})=n_s(\mathbf{r})} - \zeta = \left[-\frac{\delta T_s[\tilde{n}(\mathbf{r})]}{\delta \tilde{n}(\mathbf{r})}\right]_{\tilde{n}(\mathbf{r})=n_s(\mathbf{r})}, \tag{2.22}$$

where n_s is the ground-state density of the non-interacting system. Likewise, the minimization of the energy functional for the interacting system is given by

$$0 = \frac{\delta}{\delta \tilde{n}(\mathbf{r})}\left[\int v(\mathbf{r},\mathbf{R})\,\tilde{n}(\mathbf{r})\,d\mathbf{r} + T[\tilde{n}(\mathbf{r})] + U[\tilde{n}(\mathbf{r})] - \xi\int \tilde{n}(\mathbf{r})\,d\mathbf{r}\right]_{\tilde{n}(\mathbf{r})=n(\mathbf{r})}$$
$$= \left[v(\mathbf{r},\mathbf{R}) + \frac{\delta T_s[\tilde{n}(\mathbf{r})]}{\delta \tilde{n}(\mathbf{r})} + V_H[\tilde{n}](\mathbf{r}) + \frac{\delta E_{\text{XC}}[\tilde{n}(\mathbf{r})]}{\delta \tilde{n}(\mathbf{r})}\right]_{\tilde{n}(\mathbf{r})=n(\mathbf{r})} - \xi, \tag{2.23}$$

where the Hartree potential V_H and the Hartree energy E_H enter that are defined as

$$V_H[n](\mathbf{r}_1) = \int d\mathbf{r}_2\, u(\mathbf{r}_1,\mathbf{r}_2)\,n(\mathbf{r}_2)\quad \text{and} \tag{2.24}$$

$$E_H[n] = \frac{1}{2}\int d\mathbf{r}\, V_H[n](\mathbf{r})\,n(\mathbf{r}) = \frac{1}{2}\iint d\mathbf{r}_1\,d\mathbf{r}_2\, u(\mathbf{r}_1,\mathbf{r}_2)\,n(\mathbf{r}_1)\,n(\mathbf{r}_2). \tag{2.25}$$

Furthermore, in Eq. (2.23) the *XC functional* E_{XC} is introduced according to

$$E_{\text{XC}}[\tilde{n}(\mathbf{r})] := T[\tilde{n}(\mathbf{r})] - T_s[\tilde{n}(\mathbf{r})] + U[\tilde{n}(\mathbf{r})] - E_H[\tilde{n}], \tag{2.26}$$

which leads to the XC potential V_{XC} by means of

$$V_{\text{XC}}[n](\mathbf{r}) := \left.\frac{\delta E_{\text{XC}}[\tilde{n}(\mathbf{r})]}{\delta \tilde{n}(\mathbf{r})}\right|_{\tilde{n}(\mathbf{r})=n(\mathbf{r})}. \tag{2.27}$$

Clearly, Eq. (2.26) points out that E_{XC} should account for the difference between the kinetic energy of interacting and non-interacting electrons. It also includes all electron-electron interactions beyond the Hartree term, in particular, the quantum mechanical XC effects. Since an exact expression of this functional is unknown, a reasonable approximation is crucial.

For simplicity we disregarded two difficulties related to this approach of defining the XC functional. Firstly, v-representability of the density has been assumed, meaning that for all physically relevant densities n we require the existence of a potential v which leads to n as the ground-state density of the respective Hamiltonian. Secondly, the existence of the right-hand side of Eq. (2.22) also for all possible densities of interacting systems was implied. Neither of these problems are trivial, however, there exist respective extensions. Since this goes beyond the scope of this work we refer to specialized literature, such as Ref. [13] and the references therein. Keeping this in mind, Eqs. (2.22) and (2.23) allow defining the effective potential of the non-interacting problem (up to a constant that justifies the neglection of the Lagrange multipliers) as

$$[v_{\text{eff}}(\mathbf{r},\mathbf{R})]_{\tilde{n}(\mathbf{r})=n(\mathbf{r})} = [v(\mathbf{r},\mathbf{R}) + V_{\text{H}}[\tilde{n}](\mathbf{r}) + V_{\text{XC}}[\tilde{n}](\mathbf{r})]_{\tilde{n}(\mathbf{r})=n(\mathbf{r})}. \tag{2.28}$$

This potential leads to an effective single-particle Schrödinger equation that can be solved using a product of single-particle wave functions $\psi_l(\mathbf{r})$. The Schrödinger equations

$$\hat{H}^{\text{KS}}(\mathbf{r},\mathbf{R})\psi_l(\mathbf{r}) = \left(-\frac{\hbar^2}{2m}\Delta + v(\mathbf{r},\mathbf{R}) + V_{\text{H}}(\mathbf{r}) + V_{\text{XC}}(\mathbf{r})\right)\psi_l(\mathbf{r}) = \varepsilon_l \psi_l(\mathbf{r}) \tag{2.29}$$

for the $\psi_l(\mathbf{r})$ are referred to as the Kohn-Sham (KS) equations [4]. Their eigenstates ψ_l lead to the ground-state density, whereas the total energy of the interacting system follows from

$$E = \sum_l^{\text{occ}} \varepsilon_l - E_{\text{H}}[n] - \int d\mathbf{r}\, n(\mathbf{r}) V_{\text{XC}}[n](\mathbf{r}) + E_{\text{XC}}[n]. \tag{2.30}$$

Using the Bloch theorem [14] allows us to express the eigenstates of the single-particle Hamiltonian $\hat{H}^{\text{KS}}$ in Eq. (2.29) via a product of a plane wave and a lattice-periodic function $u_{n\mathbf{k}}(\mathbf{r})$, i.e.,

$$\langle \mathbf{r}|n\mathbf{k}\rangle = \psi_{n\mathbf{k}}(\mathbf{r}) = \frac{1}{\sqrt{\Omega}} e^{i\mathbf{k}\cdot\mathbf{r}} u_{n\mathbf{k}}(\mathbf{r}) \quad \text{with } u_{n\mathbf{k}}(\mathbf{r}+\mathbf{R}) = u_{n\mathbf{k}}(\mathbf{r}). \tag{2.31}$$

In this expression Ω denotes the volume of the crystal. Henceforth, instead of labeling the solutions of Eq. (2.29) with l, we use the two quantum numbers n and $\mathbf{k}$. The band index n counts the eigenstates that belong to the same $\mathbf{k}$ vector in the Brillouin zone (BZ).

2.2.4 Exchange and correlation

Local-density approximation and gradient corrections

In their initial paper W. Kohn and L. J. Sham [4] suggested approximations to the XC energy. Within the *local-density approximation* (LDA) the respective XC energy per electron of an homogeneous electron gas with the same density, $\varepsilon_{XC}^{hom}(n)$, is used, leading to

$$E_{XC}^{LDA}[n] = \int d\mathbf{r}\, n(\mathbf{r})\, \varepsilon_{XC}^{hom}(n)\Big|_{n=n(\mathbf{r})}. \tag{2.32}$$

An explicit expression can be given for the exchange energy of the homogeneous electron gas. Using a fit to numerical results from quantum Monte Carlo calculations, e.g. by D. Ceperley and B. Alder [15], yields a parametrization of the correlation energy.

While the LDA should work *a priori* only for systems whose density varies slowly across the average electron distance, it turned out to be very successful even when this requirement is not fulfilled. Due to this success and due to its computational simplicity, the LDA majorly contributed to the wide application of DFT. However, atomization energies of molecules or bulk materials are oftentimes found to be too large [16], while binding energies of strongly localized electrons tend to be underestimated. The overall good results for many systems have been traced back to the fact that the LDA inherently fulfills important sum rules, one example is its good reproduction of the spherical average of the XC hole, which leads to a certain error cancellation [13]. Among the deficiencies of the LDA is an artificial self-interaction [17, 18] which enters into the problem with the integral expression of the Hartree energy, Eq. (2.25). As long as the electron-electron Coulomb interaction is expressed as a sum over all electrons of the system [cf. Eq. (2.3)] the interaction of an electron with itself is explicitly excluded from the sum. In Eq. (2.25) this is obviously no longer the case. Since the integral expression for the Hartree energy has been separated from the XC functional [cf. Eqs. (2.23) and (2.26)], in principle the XC functional has to compensate for the self-interaction and the LDA does not. It turns out that the self-interaction is particularly large for the Zn $3d$ or Cd $4d$ states due to their strong localization [19]. Explicit correction schemes exist [20], however, we choose a different route in this work.

In order to improve over the LDA, including higher-order terms of an expansion of the XC functional with respect to the density of the homogeneous electron gas involves gradients of the density (gradient-expansion approximation). Unfortunately, this approach does not fulfill the aforementioned sum rules [13] and was, therefore, replaced by the also *semilocal* generalized-gradient approximation (GGA). In the context of this work the PW GGA II functional by Y. Wang and J. Perdew [16, 21, 22], also called PW91, is used for most of the ground-state calculations. Typically, the GGA slightly improves the results for ground-state properties of bulk semiconductors in comparison to the LDA, though still showing a tendency to overestimate lattice constants (by about 1 %) and underestimate cohesive energies or bulk moduli. However, the description of the localized Zn $3d$ or Cd $4d$ states is not improved with respect to the LDA [19, 23].

Orbital-dependent on-site Coulomb repulsion

Instead of applying self-interaction correction schemes, the more flexible LDA+U/GGA+U approach [24, 25], which is inspired by a Hubbard model [26], is used in this work to overcome the underbinding of the d electrons at low computational cost. Of the different formulations in the literature [24, 27, 28] we choose the representation of Dudarev et al. [25], where the additional, effective on-site Coulomb interaction is accounted for via two spherical averages of the screened Coulomb electron-electron interaction, denoted by the parameters $\bar{U}$ and $\bar{J}$. This results in an orbital-dependent modification of the LDA/GGA energy functional, i.e.,

$$E^{\text{LDA}+U/\text{GGA}+U} = E^{\text{LDA/GGA}} + \frac{\bar{U}-\bar{J}}{2} \sum_\alpha \left[\left(\sum_{d_j} n_\alpha^{d_j,d_j} \right) - \left(\sum_{d_j,d_l} n_\alpha^{d_j,d_l} n_\alpha^{d_l,d_j} \right) \right]. \quad (2.33)$$

Here, α denotes the spin of each state and $\hat{n}_\alpha^{d_j,d_l} = \hat{\psi}_\alpha^{d_l\,\dagger} \hat{\psi}_\alpha^{d_j}$ describes the operator of the single-particle density of the d states, where d_j runs over the projections of the orbital angular momentum $(-2,-1,\ldots,2)$. In Dudarev's formulation, Eq. (2.33), only one parameter U, the difference of $\bar{U}$ and $\bar{J}$, enters. Since U is not fixed a priori but strongly influences the results, we will justify its choice in detail later. The derivative of Eq. (2.33) with respect to the density [cf. Eq. (2.27)] yields the LDA+U/GGA+U contribution to the single-particle potential in the KS equations.

Hybrid functionals for exchange and correlation

When the XC functional in the KS equations (2.29) is expressed by a local or semilocal functional, their solution corresponds to approximating the many-body wave function by a single Slater determinant [29]. However, not all contributions to XC are captured by this and it turns out that generalizing the KS scheme can lead to single-particle equations with eigenvalues that resemble the excitation energies of a system much better. Such a generalization has been achieved by A. Seidl et al. [29] within the constrained-search formulation [13] of DFT. They incorporated as much information about the physics as possible by including a certain

2.2 Ground state: Density functional theory

amount of Hartree-Fock (HF) exchange, leading to a variety of *hybrid functionals* that, unfortunately, come along with a much higher computational cost compared to the LDA or the GGA.

One approach in this context, the PBE0 hybrid functional, which was independently developed by M. Ernzerhof and C. Adamo [30, 31], features an XC energy,

$$E_{XC}^{PBE0} = aE_X^{HF} + (1-a)E_X^{PBE} + E_C^{PBE}, \qquad (2.34)$$

that is based on the XC energy of the PBE-GGA functional [32] (parametrized by J. Perdew, K. Burke, and M. Ernzerhof) but contains a certain amount of HF exchange. Using arguments from the perturbation theory [33, 34] the respective value for a is fixed at $a = 1/4$.

Within this work we use the HSE functional by J. Heyd, G. Scuseria, and M. Ernzerhof [35], which exploits the fact that the exchange interaction in an insulator or semiconductor is screened and, therefore, the long-range (LR) part of the HF exchange can be truncated to reduce the computational effort. After splitting the exchange energies in Eq. (2.34) into their short-range (SR) and LR parts, the HF and the PBE LR exchange contributions tend to cancel each other [35]. Neglecting them yields

$$E_{XC}^{HSE} = aE_X^{HF,SR}(\omega) + (1-a)E_X^{PBE,SR}(\omega) + E_X^{PBE,LR}(\omega) + E_C^{PBE}. \qquad (2.35)$$

The parameter ω describes the LR/SR separation by means of the error function. It is related to a characteristic distance at which the SR interactions become negligible. For $\omega = 0$ the LR part vanishes and the SR part corresponds to the full Coulomb operator. Contrary, the functional is equivalent to PBE-GGA for $\omega \to \infty$. In the HSE functional [36–38] used in this work it holds $\omega = 0.15 a_0^{-1}$, as a good compromise between accuracy and computational cost for solids and molecules [35]. Later, they amended ω to $\omega = 0.1 a_0^{-1}$ which is the value used in the HSE06 functional [36] (see disambiguation in Ref. [39]).

2.2.5 Non-collinear spins

For the derivation of the theory so far entirely spin-paired electrons have been assumed. In order to investigate the fine structure of the one-particle spectrum, the spin-orbit interaction between the electron spin and the orbital angular momentum must be included [11]. Consequently, instead of wave functions, two-component spinors have to be taken into account in the density-functional formalism and the Pauli equation has to be solved. The electron density [cf. Eq. (2.14)] is generalized to a 2×2 spin-density *matrix*,

$$n_{\alpha\beta}(\mathbf{r}) = \left\langle \psi \left| \hat{\psi}_{\beta}^{\dagger}(\mathbf{r}) \hat{\psi}_{\alpha}(\mathbf{r}) \right| \psi \right\rangle. \tag{2.36}$$

While the sum of the diagonal components gives the electron density, $n(\mathbf{r}) = n_{\alpha\alpha} + n_{\beta\beta}$, their difference, $n_{\alpha\alpha} - n_{\beta\beta}$, describes the projection of the magnetization density onto a global quantization axis, here the z axis. The magnetization density $\mathbf{m}(\mathbf{r})$ is defined as

$$\mathbf{m}(\mathbf{r}) = \sum_{\alpha,\beta} n_{\beta\alpha}(\mathbf{r}) \cdot \boldsymbol{\sigma}_{\alpha\beta}, \tag{2.37}$$

where $\boldsymbol{\sigma} = (\sigma_x, \sigma_y, \sigma_z)$ denotes the vector of the Pauli spin matrices given by

$$\sigma_x = \begin{pmatrix} 0 & 1 \\ 1 & 0 \end{pmatrix}, \quad \sigma_y = \begin{pmatrix} 0 & -i \\ i & 0 \end{pmatrix}, \quad \sigma_z = \begin{pmatrix} 1 & 0 \\ 0 & -1 \end{pmatrix}. \tag{2.38}$$

In the most general situation the spins are non-collinear and the full $\mathbf{r}$ dependence of $n_{\alpha\beta}(\mathbf{r})$ is taken into account. In contrast, a system with collinear *spin-polarization* is described via a spin-up and a spin-down density only. In this case the expectation values of the x- and y-component of the magnetization density, as well as the off-diagonal elements of $n_{\alpha\beta}$, vanish.

The initial indication for extending the KS scheme to include spin was already given by W. Kohn and L. J. Sham [4]. In addition, in the 1970's the spin-DFT was put on a solid theoretical fundament (see Refs. [13, 40, 41] and references therein). Nevertheless, it still seems to be an open question whether a given ground state uniquely corresponds to one vector of external fields $(v(\mathbf{r}), \mathbf{B}(\mathbf{r}))$. However, a universal energy functional of $(n(\mathbf{r}), \mathbf{m}(\mathbf{r}))$ can be found and also a HK-like variational principle can be established via the constrained-search formulation of DFT [13]. The XC energy can be approximated by the expression for a spin-polarized homogeneous electron gas with the same charge and magnetization density,

$$E_{\text{XC}}^{\text{LSDA}}[n_{\alpha\beta}(\mathbf{r})] = \int d\mathbf{r}\, n(\mathbf{r})\, \varepsilon_{\text{XC}}^{\text{hom}}(n_{\alpha\alpha}(\mathbf{r})) = \int d\mathbf{r}\, n(\mathbf{r})\, \varepsilon_{\text{XC}}^{\text{hom}}(n(\mathbf{r}), m_z(\mathbf{r})). \tag{2.39}$$

In this work the non-collinear spin-DFT [42] is used to evaluate the spin-orbit interaction term that enters the KS Hamiltonian, i.e.,

$$H_{\text{SO}}(\mathbf{r}) = \frac{\hbar}{4m^2 c^2} \boldsymbol{\sigma} \cdot \left[\nabla V(\mathbf{r}) \times \frac{\hbar}{i} \nabla \right]. \tag{2.40}$$

Furthermore, we use collinear spin-polarized DFT to describe systems that have, e.g. due to a point defect, an odd number of electrons, as will be discussed later.

2.3 One-particle excitations

In the preceding section a formalism was introduced that, given a reasonable approximation for the XC functional, is suitable for the calculation of the ground-state density and total energy of an interacting many-electron system. One central goal of this work is the computation of electronic band structures, i.e., the **k**-dependent one-particle excitation energies. Oftentimes the KS eigenvalues $\varepsilon_{n\mathbf{k}}$ [cf. Eq. (2.29)] are abused by being interpreted as excitation energies. This is fundamentally and conceptually very problematic, since they are (*a priori* physically meaningless) Lagrange multipliers in the KS formalism. However, it turned out empirically that using the $\varepsilon_{n\mathbf{k}}$ as excitation energies is partially successful, especially for the description of the valence bands (VBs) of bulk semiconductors without d electrons.

Nevertheless, even for such comparably simple materials the deficiencies become manifest when gaps turn out to be too small ("LDA gap-underestimation") or wrong band dispersions occur, e.g., VB widths are underestimated [43]. The *measurement* of such quantities involves the removal of an electron or a hole from the system which, obviously, is then no longer in its ground state. Since such an *excitation* is not a small perturbation, we have to take the reaction of the system's electrons into account. In 1965 L. Hedin introduced a system of coupled integro-differential equations [5, 6] to adequately treat the excitations of electrons. Since an exact solution of the full system of Hedin's equations is impossible, approximations based on the Green's function approach are introduced.

2.3.1 Green's function and equation of motion

Using the statistical operator of the grand canonical ensemble for the interacting many-electron system, the single-particle Green's function can be defined by means of a thermodynamic average, indicated by $\langle \ldots \rangle$. In the following we neglect the spin quantum numbers and restrict our considerations to spin-paired electrons only. With the time-ordering operator $\hat{\mathcal{T}}$, the creator $\hat{\psi}^\dagger(\mathbf{r},t)$, and the annihilator $\hat{\psi}(\mathbf{r},t)$ the Green's function reads [44]

$$G\left(\mathbf{r}t,\mathbf{r}'t'\right) = G\left(\mathbf{r}\mathbf{r}', t-t'\right) = \frac{1}{i\hbar} \left\langle \hat{\mathcal{T}} \left\{ \hat{\psi}(\mathbf{r},t)\, \hat{\psi}^\dagger(\mathbf{r}',t') \right\} \right\rangle. \qquad (2.41)$$

For times $t > t'$ ($t < t'$) we can interpret this expression as the probability amplitude to find an electron (hole) at time t and position $\mathbf{r}$ that evolved from an electron (hole) which was created at time t' and position $\mathbf{r}'$. The Green's function can be related to an experimentally accessible quantity, the spectral weight function [45], which, in the case of non-interacting electrons, is composed of delta-function-like peaks located at the excitation energies of the system. These peaks are infinitely sharp due to the infinite lifetime of the excitations of non-interacting electrons. In the presence of an interaction, the energetic positions of the spectral function's peaks change and broadening occurs due to the finite lifetimes. Nevertheless, for long-living excitations of the system we still expect distinct peak structures. The rest of the spectral weight is attributed to satellite structures. In the *quasiparticle (QP) picture* an excitation of the interacting system is interpreted as a single-particle excitation at the new energetic position of the respective peak maximum of the spectral function. This approximation works well for weakly-correlated systems [45], i.e., when the problem of interacting particles can be transformed to weakly interacting QPs.

Ultimately, finding the excitation energies is equivalent to finding the poles of the Green's function in the complex energy plane. To calculate the Green's function, the Heisenberg picture, in which the operators themselves are time-dependent, is used. The equation of motion for the field operators is then

$$i\hbar \frac{\partial}{\partial t} \hat{\Psi}(\mathbf{r},t) = [\hat{\Psi}(\mathbf{r},t), \hat{H}(\mathbf{r})] \qquad \text{with } \hat{\Psi} \in \{\hat{\psi}, \hat{\psi}^\dagger\}. \qquad (2.42)$$

Using the commutator relations, Eqs. (2.5) and (2.6), together with the time derivative of the Green's function one obtains

$$\delta(\mathbf{r}-\mathbf{r}')\delta(t-t') = \left[i\hbar \frac{\partial}{\partial t} + \frac{\hbar^2}{2m}\Delta_\mathbf{r} - v(\mathbf{r},\mathbf{R})\right] G(\mathbf{rr}',t-t') \qquad (2.43)$$
$$- \frac{1}{i\hbar} \int d\mathbf{r}'' u(\mathbf{r},\mathbf{r}'') \left\langle \{\hat{\psi}^\dagger(\mathbf{r}'',t)\,\hat{\psi}(\mathbf{r}'',t)\,\hat{\mathcal{T}}\hat{\psi}(\mathbf{r},t)\,\hat{\psi}^\dagger(\mathbf{r}',t')\} \right\rangle.$$

The second factor of the integrand strongly resembles the definition of the single-particle Green's function [cf. Eq. (2.41)], except that in Eq. (2.43) four field operators appear. In fact, this term is the *two-particle* Green's function [44] which obeys its own equation of motion. Repeated application of these steps leads to continuously increasing orders of Green's functions and, therefore, an entire hierarchy of equations. We obtain from Eq. (2.43)

$$\delta(\mathbf{r}-\mathbf{r}')\delta(t-t') = \left[i\hbar \frac{\partial}{\partial t} + \frac{\hbar^2}{2m}\Delta_\mathbf{r} - v(\mathbf{r},\mathbf{R})\right] G(\mathbf{rr}',t-t') \qquad (2.44)$$
$$+ i\hbar \iint d\mathbf{r}'' dt'' u(\mathbf{r},\mathbf{r}'') \delta(t-t'') G(\mathbf{rr}'',t-t'';\mathbf{r}'\mathbf{r}'',t'-t''+i\gamma),$$

with an infinitesimally small γ to take the time ordering into account.

2.3.2 The electronic self-energy

After separating the Hartree term from the electron-electron interaction on the right-hand side of Eq. (2.44) the *self-energy* Σ (containing all XC effects) is introduced, which leads to

$$\delta(\mathbf{r}-\mathbf{r}')\delta(t-t') = \left[i\hbar\frac{\partial}{\partial t} + \frac{\hbar^2}{2m}\Delta_\mathbf{r} - v(\mathbf{r},\mathbf{R}) - V_\mathrm{H}(\mathbf{r})\right] G\left(\mathbf{rr}',t-t'\right) \qquad (2.45)$$
$$- \iint \mathrm{d}\mathbf{r}'' \, \mathrm{d}t'' \Sigma\left(\mathbf{rr}'',t-t''\right) G\left(\mathbf{r}''\mathbf{r}',t''-t'\right).$$

Within the QP approximation, i.e., assuming only one pole with full spectral weight and, hence, neglecting the satellite structures, this equation can be formally solved by the spectral representation of the Green's function

$$G\left(\mathbf{rr}',\omega\right) = \sum_{n\mathbf{k}} \frac{\psi_{n\mathbf{k}}^{\mathrm{QP}}(\mathbf{r}) \, \psi_{n\mathbf{k}}^{\mathrm{QP}\,*}(\mathbf{r}')}{\hbar\omega - \varepsilon_{n\mathbf{k}}^{\mathrm{QP}}}, \qquad (2.46)$$

where the $\varepsilon_{n\mathbf{k}}^{\mathrm{QP}}$ and $\psi_{n\mathbf{k}}^{\mathrm{QP}}(\mathbf{r})$ are solutions of the QP equation

$$\left[-\frac{\hbar^2}{2m}\Delta_\mathbf{r} + v(\mathbf{r},\mathbf{R}) + V_\mathrm{H}(\mathbf{r})\right] \psi_{n\mathbf{k}}^{\mathrm{QP}}(\mathbf{r}) + \int \mathrm{d}\mathbf{r}' \Sigma(\mathbf{rr}', \varepsilon_{n\mathbf{k}}^{\mathrm{QP}}/\hbar) \psi_{n\mathbf{k}}^{\mathrm{QP}}(\mathbf{r}') = \varepsilon_{n\mathbf{k}}^{\mathrm{QP}} \psi_{n\mathbf{k}}^{\mathrm{QP}}(\mathbf{r}). \qquad (2.47)$$

Even though this expression is reminiscent of the KS equation [cf. Eq. (2.29)], it is much more complicated because Σ is generally a non-local, non-Hermitian, and energy-dependent operator.

Obviously, approximations are necessary for practical applications. It was expected that an expansion with respect to the bare Coulomb potential would converge poorly, while expanding Σ in terms of the *screened* Coulomb interaction

$$W\left(\mathbf{rr}',t-t'\right) = \iint \mathrm{d}\mathbf{r}'' \, \mathrm{d}t'' \, \varepsilon^{-1}\left(\mathbf{rr}'',t-t''\right) u(\mathbf{r}'',\mathbf{r}') \delta(t''-t') \qquad (2.48)$$

should be much more successful [5, 6]. The dynamical and non-local reaction of the system to a single-particle excitation is included in W, Eq. (2.48), via the screening of the Coulomb potential within linear response by means of the inverse microscopic dielectric function (DF) ε^{-1} [44] of independent particles. Truncating the expansion after the first term and, hence, neglecting vertex corrections, gives Hedin's GW approximation [5, 6, 46] of the self-energy

$$\Sigma^{GW}\left(\mathbf{rr}',t-t'\right) \equiv i\hbar \, G\left(\mathbf{rr}',t-t'\right) W\left(\mathbf{r}'\mathbf{r},t'-t+i\gamma\right). \qquad (2.49)$$

It turned out that this approach works very well and has been widely adopted for a large number of materials [47–49]. Also the electronic-structure calculations in this work are based on Eq. (2.49), hence, the remaining task is to apply the best possible approximations for G and W to calculate Σ. Solving the QP equation (2.47) yields the QP energies $\varepsilon_{n\mathbf{k}}^{\text{QP}}$ and QP wave functions $\psi_{n\mathbf{k}}^{\text{QP}}(\mathbf{r})$. A comparison of Eqs. (2.29) and (2.47) indicates that, due to the XC functional, part of the self-energy is already contained in a mean-field way in the KS Hamiltonian. This fact suggests an iterative calculation of the $\varepsilon_{n\mathbf{k}}^{\text{QP}}$. We approximate them by a sum of the KS energy $\varepsilon_{n\mathbf{k}}$ and a QP shift $\Delta_{n\mathbf{k}}$ that results from the remaining self-energy effects [43], i.e.,

$$\varepsilon_{n\mathbf{k}}^{\text{QP}} = \varepsilon_{n\mathbf{k}} + \Delta_{n\mathbf{k}} = \varepsilon_{n\mathbf{k}} + Z_{n\mathbf{k}} \operatorname{Re} \langle n\mathbf{k} | \Sigma(\varepsilon_{n\mathbf{k}}/\hbar) - V_{\text{XC}} | n\mathbf{k} \rangle, \qquad (2.50)$$

$$\text{with } Z_{n\mathbf{k}} = \left(1 - \left. \frac{\partial \left(\operatorname{Re} \langle n\mathbf{k} | \Sigma(\omega) | n\mathbf{k} \rangle \right)}{\partial (\hbar \omega)} \right|_{\hbar \omega = \varepsilon_{n\mathbf{k}}} \right)^{-1}. \qquad (2.51)$$

Calculating the QP energies using only the first-order of the perturbation theory requires that the shifts $\Delta_{n\mathbf{k}}$ are small, i.e., the KS eigenvalues $\varepsilon_{n\mathbf{k}}$ are close to the $\varepsilon_{n\mathbf{k}}^{\text{QP}}$. This approach appears in the literature as the G_0W_0 approximation. Strictly speaking, its success cannot be justified *a priori*, but comparison to experimental results proves that it works well for many insulators, semiconductors, and even metals [45]. It has been shown before [47, 50], that the HSE functional, which we use as approximation to XC in this work, provides a starting electronic structure which is close to the QP results.

2.4 Two-particle excitations

In the preceding section we discussed those excitations which caused electrons or holes to be removed from the system. When the relevant excitation energies are lower, e.g., in optical measurements, the electron and the hole are not moved apart during the process and remain in the system. Moreover, due to their opposite charge these particles attract each other via the Coulomb interaction and form an electron-hole pair. This two-particle excitation of the system corresponds to the creation of a bosonic QP called an *exciton*.

In principle, electron-hole pairs can be studied using Hedin's system of equations [5, 6]. In this section, approximations and theoretical concepts are introduced and discussed that are suitable for achieving the calculation of the optical properties of a system, such as its DF or its absorption coefficient, taking excitonic effects into account. The DF especially plays a fundamental role since all linear optical properties can be derived from it.

2.4.1 Bethe-Salpeter equation

The optical response of an interacting many-electron system can be related to its polarization function P. Moreover, P is connected to the DF of the system via

$$\varepsilon(11') = \delta(1-1') - \int d2\, u(1-2) P(21'). \tag{2.52}$$

Here and in the following we use the short-hand notation

$$(1) := (\mathbf{r}_1, t_1), \qquad (1^+) := (\mathbf{r}_1, t_1 - i\gamma), \qquad u(1-2) := u(\mathbf{r}_1, \mathbf{r}_2)\, \delta(t_1 - t_2).$$

A Bethe-Salpeter equation (BSE) exists for the four-point polarization function P [51],

$$P(11', 22') = P_0(11', 22') + \iiiint d3\, d4\, d5\, d6\, P_0(11', 43)\, \Xi(34, 65)\, P(56, 22'), \tag{2.53}$$

with a kernel that determines the electron-hole interaction,

$$\Xi(12, 34) = -\frac{1}{i\hbar} \frac{\delta \Sigma(12)}{\delta G(43)}, \tag{2.54}$$

and the polarization function of independent QPs P_0 [cf. Eq. (2.58)]. Employing Hedin's GW approximation of the self-energy Σ, Eq. (2.49), leads to two terms as a result of the variation in Eq. (2.54), one being proportional to the screened Coulomb potential W, Eq. (2.48). While this term describes the screened Coulomb attraction between the electron and the hole, the other one (the variation of W with respect to the Green's function) is usually neglected, which is well-justified for bulk materials with dispersive energy bands [52].

To relate the microscopic DF, Eq. (2.52), to the macroscopic dielectric tensor $\varepsilon_M(\omega)$, the microscopic reaction of the system to a macroscopic perturbation must be included. Adler [53] and Wiser [54] independently showed that this can be achieved by means of the expression

$$\varepsilon_M(\hat{\mathbf{q}}, \omega) = \hat{\mathbf{q}} \cdot \hat{\varepsilon}_M(\omega) \cdot \hat{\mathbf{q}} = \lim_{q \to 0} \frac{1}{\varepsilon^{-1}(\mathbf{q}+\mathbf{G}, \mathbf{q}+\mathbf{G}'; \omega)}\bigg|_{\mathbf{G}=\mathbf{G}'=0}, \tag{2.55}$$

where $\mathbf{G}$ and $\mathbf{G}'$ are vectors of the reciprocal lattice. Furthermore, we indicate with $\hat{\mathbf{q}} := \frac{\mathbf{q}}{|\mathbf{q}|}$ the direction of the (in the optical limit vanishing) wave vector $\mathbf{q}$ of a photon with the energy $\hbar\omega$. Since the inversion (2.55) of the microscopic DF for all frequencies is computationally expensive, the local-field effects are taken into account by means of a short-range Coulomb potential [55]

$$\bar{u}(\mathbf{q}+\mathbf{G}) = \begin{cases} u(\mathbf{q}+\mathbf{G}) & \mathbf{G} \neq 0 \\ 0 & \mathbf{G} = 0 \end{cases}. \tag{2.56}$$

The term $\bar{u}$ is equal to the Coulomb potential without its long-range Fourier component. It accounts for the local-field effects via a bare Coulomb exchange term [51, 56] in the resulting BSE kernel

$$\Xi(34,65) = \delta(3-4)\,\delta(5-6)\,\bar{u}(35) - \delta(3-5)\,\delta(4-6)\,W(3^+4). \tag{2.57}$$

Finally, with the kernel (2.57) the BSE for the polarization function, Eq. (2.53), governs the macroscopic DF including excitonic and local-field effects by means of a relation similar to Eq. (2.52). Due to the frequency dependence of the screened Coulomb potential also the kernel, Eq. (2.57), is still fully frequency dependent. We restrict ourselves to static screening only, leading to $W(12) \approx W(\mathbf{r}_1\mathbf{r}_2)\delta(t_1 - t_2)$. Only in this case a closed equation for the polarization function depending merely on one frequency exists [51].

2.4.2 Excitonic Hamiltonian

To achieve a solution of the BSE, Eq. (2.53), it is expressed in terms of Bloch states. We use the Bloch representation of P_0,

$$P_0(\lambda_1\lambda_1',\lambda_2\lambda_2';\omega) = \frac{n_{\lambda_1} - n_{\lambda_1'}}{\varepsilon_{\lambda_1}^{QP} - \varepsilon_{\lambda_1'}^{QP} - \hbar(\omega+i\gamma)} \delta_{\lambda_1\lambda_2'} \delta_{\lambda_2\lambda_1'}, \tag{2.58}$$

where λ cumulates the quantum numbers n and $\mathbf{k}$ of the Bloch states and the n_λ denote their occupation numbers. We obtain for Eq. (2.53)

$$P(\lambda_1\lambda_1',\lambda_2\lambda_2';\omega) = \frac{n_{\lambda_1} - n_{\lambda_1'}}{\varepsilon_{\lambda_1}^{QP} - \varepsilon_{\lambda_1'}^{QP} - \hbar(\omega+i\gamma)} \left(\delta_{\lambda_1\lambda_2'}\delta_{\lambda_2\lambda_1'} + \sum_{\kappa\kappa'} \Xi(\lambda_1\lambda_1',\kappa'\kappa) P(\kappa\kappa',\lambda_2\lambda_2';\omega) \right). \tag{2.59}$$

A trivial solution for Eq. (2.59) is given by $P(\lambda_1\lambda_1',\lambda_2\lambda_2';\omega) \equiv 0$ for a vanishing difference $n_{\lambda_1} - n_{\lambda_1'} = 0$. These cases are excluded in the following [57], in which case the problem can be rewritten as

$$\sum_{\kappa,\kappa'} \left\{ H(\lambda_1\lambda_1',\kappa\kappa') - \hbar(\omega+i\gamma)\delta_{\lambda_1\kappa}\delta_{\lambda_1'\kappa'} \right\} P(\kappa\kappa',\lambda_2\lambda_2';\omega) = \left(n_{\lambda_1} - n_{\lambda_1'}\right)\delta_{\lambda_1\lambda_2'}\delta_{\lambda_1'\lambda_2} \tag{2.60}$$

$$\text{with } H(\lambda_1\lambda_1',\lambda_2\lambda_2') = \left(\varepsilon_{\lambda_1}^{QP} - \varepsilon_{\lambda_1'}^{QP}\right)\delta_{\lambda_1\lambda_2}\delta_{\lambda_1'\lambda_2'} - \left(n_{\lambda_1} - n_{\lambda_1'}\right)\Xi(\lambda_1\lambda_1',\lambda_2'\lambda_2). \tag{2.61}$$

2.4 Two-particle excitations

The operator H in Eq. (2.61) is only Hermitian for integer occupation numbers, i.e., when their difference $n_{\lambda_1} - n_{\lambda'_1}$ equals 1 or -1, as it is the case for undoped semiconductors at $T = 0$ K. Restricting our considerations to electron-hole excitations, i.e., processes that conserve the particle number, allows the specifying of λ by c for conduction bands (CBs) and v for VBs. Within the Tamm-Dancoff approximation that neglects the coupling of the resonant and the anti-resonant parts of the *excitonic Hamiltonian H* we obtain the eigenvalue problem

$$H(cv\mathbf{k}, c'v'\mathbf{k}') = \left(\varepsilon_{c\mathbf{k}}^{\text{QP}} - \varepsilon_{v\mathbf{k}}^{\text{QP}}\right)\delta_{cc'}\delta_{vv'}\delta_{\mathbf{k}\mathbf{k}'} - \Xi(cv\mathbf{k}, c'v'\mathbf{k}'), \tag{2.62}$$

$$\sum_{c',v',\mathbf{k}'} H(cv\mathbf{k}, c'v'\mathbf{k}') A_\Lambda(c'v'\mathbf{k}') = E_\Lambda A_\Lambda(cv\mathbf{k}), \tag{2.63}$$

for the resonant part [55]. The respective eigenstates A_Λ and eigenvalues E_Λ can be used to calculate the optical properties of the system [58–60]. More specifically, the E_Λ are the optical excitation energies with excitonic and local-field effects included, whereas the A_Λ can be related to the optical oscillator strength $\mathbf{F} = (F_x, F_y, F_z)$ of the corresponding transitions via [61]

$$\mathbf{F}_\Lambda = \frac{2}{m}\left|\sum_{cv\mathbf{k}} A_\Lambda^*(cv\mathbf{k})\frac{\langle c\mathbf{k}|\mathbf{p}|v\mathbf{k}\rangle}{\varepsilon_{c\mathbf{k}} - \varepsilon_{v\mathbf{k}}}\right|^2 E_\Lambda. \tag{2.64}$$

In Eq. (2.64) the optical transition-matrix elements of non-interacting electron-hole pairs enter,

$$M_j(cv\mathbf{k}) = \frac{e\hbar}{im}\frac{\langle c\mathbf{k}|p_j|v\mathbf{k}\rangle}{\varepsilon_{c\mathbf{k}} - \varepsilon_{v\mathbf{k}}}, \tag{2.65}$$

where p_j is the jth component of the momentum operator. Expression (2.64) demonstrates how *all* transitions between non-interacting electron and hole states contribute to the oscillator strength of *one* interacting electron-hole pair Λ. Moreover, due to its vector character, $\mathbf{F}$ allows access to the polarization dependence imposed on the optical properties due to the symmetry constraints of the crystal lattice. For the oxides studied in this work, the number of independent components of the dielectric tensor [cf. Eq. (2.55)] is reduced to only one (two) in the case of the rocksalt (wurtzite) crystal structure. Therefore, for cubic rocksalt (*rs*) crystals it holds $\varepsilon(\omega) = \varepsilon_{xx}(\omega) = \varepsilon_{yy}(\omega) = \varepsilon_{zz}(\omega)$ [53]. For hexagonal wurtzite (*wz*) crystals $\varepsilon_{xx}(\omega) = \varepsilon_{yy}(\omega)$ correspond to ordinary light polarization ($\mathbf{E}$ perpendicular ($\perp$) to the crystals' c axis) and $\varepsilon_{zz}(\omega)$ to extraordinary light polarization ($\mathbf{E}$ parallel ($\parallel$) to c).

For real systems oftentimes the direct diagonalization of the excitonic Hamiltonian is numerically too demanding. Its rank N is fixed by the number of electron-hole pair states $N = N_v \cdot N_c \cdot N_{\text{KP}}$, where N_v counts all VBs, N_c all CBs, and N_{KP} all $\mathbf{k}$ points. The *lowest* eigenstates and eigenvalues of H, Eq. (2.62), can be accessed at significantly reduced computational cost via an iterative-diagonalization scheme [61] which only scales quadratically with the rank N and, hence, allows us to study the lowest optical transitions for matrices with $N \approx 100,000$.

2.4.3 Macroscopic dielectric function

In addition to the calculation of single excitations of the system, we proceed now with the derivation of an expression for the DF. Restricting the treatment to interband transitions and using the Bloch representation of the polarization function we obtain for Eq. (2.52)

$$\varepsilon_M(\hat{\mathbf{q}}, \omega) = 1 - \frac{2e^2\hbar^2}{\Omega \varepsilon_0 m^2} \sum_{cv\mathbf{k}, c'v'\mathbf{k}'} \frac{\langle c\mathbf{k}|\hat{\mathbf{q}}\mathbf{p}|v\mathbf{k}\rangle^* \langle c'\mathbf{k}'|\hat{\mathbf{q}}\mathbf{p}|v'\mathbf{k}'\rangle}{\varepsilon_{c\mathbf{k}} - \varepsilon_{v\mathbf{k}}} P(cv\mathbf{k}, c'v'\mathbf{k}'; \omega). \qquad (2.66)$$

With the solution of the eigenvalue problem (2.63) we can rewrite Eq. (2.66) as

$$\varepsilon_M(\hat{\mathbf{q}}, \omega) = 1 + \frac{2e^2\hbar^2}{\Omega \varepsilon_0 m^2} \sum_{\Lambda} \left| \sum_{cv\mathbf{k}} \frac{\langle c\mathbf{k}|\hat{\mathbf{q}}\mathbf{p}|v\mathbf{k}\rangle^*}{\varepsilon_{c\mathbf{k}} - \varepsilon_{v\mathbf{k}}} A_\Lambda(cv\mathbf{k}) \right|^2 \left(\frac{1}{E_\Lambda - \hbar(\omega + i\gamma)} + \frac{1}{E_\Lambda + \hbar(\omega + i\gamma)} \right). \qquad (2.67)$$

If we knew the respective E_Λ and A_Λ, it would now be possible to calculate the macroscopic DF, including excitonic and local-field effects. Fortunately, we can again bypass the prohibitively expensive direct diagonalization of Eq. (2.62) by transforming the problem of calculating the DF into an initial-value problem [62, 63] which can be treated using an efficient time-evolution scheme. This approach also features a quadratic scaling behavior with the rank N of the excitonic Hamiltonian and, therefore, enables us to treat large matrices as well.

2.4.4 Screening in heavily doped materials

For studying the macroscopic DF of semiconductors with filled VBs that are well-separated from the empty CBs, only interband transitions need to be taken into account. Now we assume that, e.g. due to heavy doping, the lowest CB can be partially filled and that its shape remains unchanged. We are interested in the *intraband* contributions to the DF, that become important in this case [64]. All other screening effects (electronic or lattice contributions) may be described by an effective static dielectric constant ε_{eff}. To incorporate the impact of the free electrons on the screening into the static kernel of the BSE we decompose the static

2.4 Two-particle excitations

DF according to

$$\varepsilon(\mathbf{q}) = \varepsilon_{\text{eff}} + \varepsilon'_{\text{intra}}(\mathbf{q}) = \varepsilon_{\text{eff}}\left(1 + \frac{\varepsilon'_{\text{intra}}(\mathbf{q})}{\varepsilon_{\text{eff}}}\right) = \varepsilon_{\text{eff}}\varepsilon_{\text{intra}}(\mathbf{q}). \quad (2.68)$$

We assume that the additional electrons in the material form a degenerate electron gas in the lowest CB. Therefore, we evaluate the DF for finite wave vectors $\mathbf{q}$ and obtain for $\varepsilon_{\text{intra}}(\mathbf{q})$ the DF of *free* electrons [64]

$$\varepsilon_{\text{intra}}(\mathbf{q}) = 1 + 2\frac{e^2}{\varepsilon_0\varepsilon_{\text{eff}}\Omega q^2}\sum_{\mathbf{k}}\frac{n_{c\mathbf{k}}}{\varepsilon_{c\mathbf{k}}^{\text{QP}} - \varepsilon_{c\mathbf{k}+\mathbf{q}}^{\text{QP}}}. \quad (2.69)$$

Rewriting the sum (2.69) as an integral and carrying out the integration over the CB states, that are occupied by the degenerate electrons and that we assume to be isotropic and parabolic (effective mass m_c), leads to the static Lindhard DF

$$\varepsilon_{\text{intra}}(q) = 1 + \frac{e^2}{\varepsilon_0\varepsilon_{\text{eff}}q^2}\frac{2n_c}{3\varepsilon_F}\left[\frac{1}{2} + \frac{4k_F^2 - q^2}{8k_F q}\ln\left|\frac{2k_F + q}{2k_F - q}\right|\right]. \quad (2.70)$$

The density of the free electrons n_c in the band c is related to their Fermi vector $k_F = \sqrt[3]{3\pi^2 n_c}$. With $\varepsilon_F = \hbar^2 k_F^2/(2m_c)$ we denote the Fermi energy with respect to the CB minimum (CBM). It turns out that the impact on the screening is especially large for small q (large distances), whereas it vanishes for large q (small distances) [64]. This indicates that the free electrons (FE) efficiently screen outside a certain screening length. More specifically, introducing the Thomas-Fermi (TF) wave vector q_{TF}, the limit of Eq. (2.68) for vanishing q follows as

$$\varepsilon(q) = \varepsilon_{\text{eff}}\varepsilon_{\text{intra}}(q) \approx \varepsilon_{\text{eff}}\left(1 + \frac{q_{\text{TF}}^2}{q^2}\right) \quad \text{with } q_{\text{TF}} = \sqrt{\frac{3n_c e^2}{2\varepsilon_0\varepsilon_{\text{eff}}\varepsilon_F}}. \quad (2.71)$$

Via $\varepsilon(q)$, Eq. (2.71), we describe the reduction of an external perturbing potential $W_{\text{ext}}(\mathbf{q})$ due to the reaction of the electrons to the presence of W_{ext}, leading to

$$W_{\text{FE}}(\mathbf{r}) = \sum_{\mathbf{q}}\frac{W_{\text{ext}}(\mathbf{q})}{\varepsilon(q)}e^{i\mathbf{q}\mathbf{r}}, \quad (2.72)$$

where $\mathbf{q}$ runs over all reciprocal lattice vectors. For a point charge of $-e$ in the medium as an external perturbation with the potential $W_{\text{ext}}(q) = e^2/(\varepsilon_0\Omega q^2)$ we obtain

$$W_{\text{FE}}(\mathbf{r}) = \frac{e^2}{4\pi\varepsilon_0\varepsilon_{\text{eff}}r}e^{-q_{\text{TF}}r}. \quad (2.73)$$

Expression (2.73) corresponds to the real-space representation of the short-range Yukawa potential. The preceding considerations showed that the degenerate electron gas in the lowest CB is the reason for a strong, *additional* screening. Consequently, when free electrons are present, the screened electron-hole Coulomb interaction W [cf. Eq. (2.48)] becomes a Yukawa-type potential which has, as a short-range potential, only a finite number of bound states [65]. Assuming a parabolic VB with the effective mass m_v allows to estimate the Bohr radius of the exciton from

$$a_{\text{B,exc}} = a_\text{B} \frac{\varepsilon_{\text{eff}} m}{\mu} \quad \text{with } \mu = \frac{m_c m_v}{m_c + m_v}, \quad (2.74)$$

where μ is the reduced mass of the electron-hole pair and a_B is the Bohr radius of the hydrogen atom. In numerical simulations [66] it has been found that for $q_{\text{TF}} a_{\text{B,exc}} > 1.19$ no bound electron-hole pair states exist anymore for the Yukawa potential, which then leads to the Mott density n_M,

$$n_\text{M} = \left(\frac{1.19 \cdot \mu}{2\sqrt{m_c m \varepsilon_{\text{eff}}}}\right)^6 \frac{\pi}{3} \left(\frac{1}{a_\text{B}}\right)^3, \quad (2.75)$$

as the free-carrier density for which an unbinding of electrons and holes, the so-called *Mott transition* of the exciton, occurs.

2.4.5 Semiconductor Bloch equations

In addition to the full *ab-initio* approach to treat electron-hole pair excitations we want to begin our investigation of the excitonic effects in heavily doped systems in Chapter 8 by studying the underlying physics via a two-band model [65]. Therefore, we start with the Bloch representation of the density matrix for a lattice with translational symmetry

$$\left\langle \hat{a}^\dagger_{n_2 \mathbf{k}_2}(t) \hat{a}_{n_1 \mathbf{k}_1}(t) \right\rangle = \delta_{\mathbf{k}_1,\mathbf{k}_2} n_{n_1 \mathbf{k}_1, n_2 \mathbf{k}_1}(t), \quad (2.76)$$

where the $\hat{a}^\dagger_{n\mathbf{k}}$ and $\hat{a}_{n\mathbf{k}}$ are the creators and annihilators for an electron in the QP state described by n and $\mathbf{k}$. The translational symmetry ensures the diagonality of Eq. (2.76) with respect to $\mathbf{k}$ [65] and we can write this matrix for one VB v and one CB c as

$$N_\mathbf{k}(t) = \begin{pmatrix} n_{c\mathbf{k}}(t) & n_{c\mathbf{k},v\mathbf{k}}(t) \\ n^*_{v\mathbf{k},c\mathbf{k}}(t) & n_{v\mathbf{k}}(t) \end{pmatrix}. \quad (2.77)$$

To probe the excitations of the system, a weak external electric pulse of the form $\mathbf{E}(t) = \underline{\mathbf{E}}(t) + \underline{\mathbf{E}}^*(t)$, with $\underline{\mathbf{E}}(t) = \underline{\mathbf{E}}_0 e^{-i\omega t}$, is introduced. It couples to the system by means of a dipole-matrix

2.4 Two-particle excitations

element M which, for simplicity, we choose to be real and constant here, i.e., we neglect the $\mathbf{k}$- and direction-dependence of Eq. (2.65). The effective electron-hole pair Hamiltonian then reads

$$H_{\mathbf{k}}(t) = \begin{pmatrix} \varepsilon_{c\mathbf{k}}^{QP} & -M\underline{E}(t) \\ -M^*\underline{E}^*(t) & \varepsilon_{v\mathbf{k}}^{QP} \end{pmatrix} - \frac{1}{\Omega'}\sum_{\mathbf{k}'} W(\mathbf{k}-\mathbf{k}') \left[N_{\mathbf{k}'}(t) - N_{\mathbf{k}'}^{(0)} \right], \quad (2.78)$$

with the initial density matrix $N_{\mathbf{k}}^{(0)}$ for $\underline{E}(t) \equiv 0$ and the matrix element $W(\mathbf{k}-\mathbf{k}')$ of the electron-hole interaction. In Eq. (2.78), Ω' determines the volume for which the $\mathbf{k}$ summation is carried out. Using this two-band model, we compare the statically screened Coulomb potential, i.e. $W(\mathbf{k}-\mathbf{k}') = e^2/(\varepsilon_0\varepsilon_{\mathrm{eff}}|\mathbf{k}-\mathbf{k}'|^2)$, as approximation for the electron-hole interaction in the absence of a degenerate electron gas in the lowest CB, to the Yukawa potential (cf. Section 2.4.4).

The solutions to the problem (2.78) are obtained via the equations of motion for the density matrix [cf. Eq. (2.76)], also known as semiconductor Bloch equations [65, 67],

$$i\hbar\frac{\partial}{\partial t}N_{\mathbf{k}}(t) = [H_{\mathbf{k}}(t), N_{\mathbf{k}}(t)]. \quad (2.79)$$

To study the linear optical properties, only terms that are linear in $\underline{E}(t)$ are kept and both $n_{c\mathbf{k}}$ and $n_{v\mathbf{k}}$ remain constant under the influence of the excitation. It then follows from Eq. (2.79) for the time evolution of the electron-hole amplitude $n_{c\mathbf{k},v\mathbf{k}}$ that

$$i\hbar\frac{\partial}{\partial t}n_{c\mathbf{k},v\mathbf{k}}(t) = \left[\varepsilon_{c\mathbf{k}}^{QP} - \varepsilon_{v\mathbf{k}}^{QP} \right] n_{c\mathbf{k},v\mathbf{k}}(t) - M\underline{E}(t)[n_{v\mathbf{k}} - n_{c\mathbf{k}}] - [n_{v\mathbf{k}} - n_{c\mathbf{k}}]\sum_{\mathbf{k}'} W(\mathbf{k}-\mathbf{k}') n_{c\mathbf{k}',v\mathbf{k}'}(t), \quad (2.80)$$

where the occupation numbers of the VB ($n_{v\mathbf{k}}$) and the CB ($n_{c\mathbf{k}}$) enter explicitly. The first term on the right-hand side of Eq. (2.80) describes the optical transition energies of non-interacting QPs. The second term describes the external pulse that probes the excitation energies of the system and finally, with the last term, the interaction of the electron and the hole enters. The term $n_{c\mathbf{k},v\mathbf{k}}$ is, by definition [cf. Eq. (2.76)], related to the optical transition probability. Therefore, the optical polarization P_M of the system can be calculated from $n_{c\mathbf{k},v\mathbf{k}}$ via the expression

$$P_M(t) = \underline{P}_M(t) + \underline{P}_M^*(t) \quad \text{where} \quad \underline{P}_M(t) = M^*\frac{1}{\Omega'}\sum_{\mathbf{k}} n_{c\mathbf{k},v\mathbf{k}}(t). \quad (2.81)$$

Within the rotating wave approximation $n_{c\mathbf{k},v\mathbf{k}}(t)$ and $\underline{P}_M(t)$ have only positive frequency components [65]. In the *linear response regime* the response functions themselves are independent of the external field and in this case one obtains for the absorption coefficient

$$\alpha(\omega) \propto \mathrm{Im}\left[\frac{P_\mathrm{M}(\omega)}{|\mathbf{E}(\omega)|}\right], \quad (2.82)$$

using the Fourier transforms of the macroscopic polarization ($P_\mathrm{M}(\omega)$) and of the electric field of the external pulse ($\mathbf{E}(\omega)$). We want to remark, that the semiconductor Bloch equations are not the only way to access the optical polarization of the system. Moreover, the equation of motion for the $n_{c\mathbf{k},v\mathbf{k}}$ can be integrated within Dirac time-dependent perturbation theory. Keeping only the first order of the expansion, i.e., terms that are linear in the electric field, also leads to the same result for the polarization and, therefore, the optical properties.

2.5 Alloy statistics and thermodynamics

An important goal of materials science is to go beyond merely understanding matter and achieve the *adjusting* of certain structural or electronic characteristics of a compound in correspondence to particular applications. Within this work we investigate the possibility of tuning the electronic structure, which is of practical relevance, for instance, in the context of band-gap tailoring. Since the fundamental gaps of MgO, ZnO, and CdO span a range of several eV, these oxides seem to be particularly interesting candidates for alloying.

Therefore, in this section, basic theoretical concepts for the treatment of pseudobinary alloys by means of *ab-initio* calculations are introduced. Using a cluster expansion, the behavior of the macroscopic system is related to properties of a set of elementary clusters that constitute the alloy. In addition, the connection to the related thermodynamic properties will be established. Our considerations are based on the theoretical approaches that have been developed over the last 20 years to describe isostructural, pseudobinary alloys of the type $A_xB_{1-x}C$ [68–71].

2.5.1 Cluster expansion

Within our description the isostructural, pseudobinary alloys $A_xB_{1-x}C$ consist of N atoms of type C on the anion sublattice and N atoms of type A or B on the cation sublattice. This macroscopic alloy is divided into an ensemble of M clusters consisting of $2n$ atoms each (n anions and n cations). The total number of cations or anions is given by $N = nM$.

From combinatorics it follows that, for a given crystal structure, there are 2^n different possibilities of arranging A- or B-type atoms on the n cation sites of one cluster while the occupation of the anion sublattice is fixed. Due to the symmetry of the crystal lattice, the clusters can be grouped in $J+1$ different classes, with J depending on the actual crystal structure. Each class j ($j = 0,\ldots,J$) contains g_j clusters of the same total energy ε_j, with the degeneracy factors g_j fulfilling the relation $\sum_j g_j = 2^n$.

To each macroscopic alloy one can assign a cluster set $\{M_0, M_1, \ldots, M_J\}$ which describes how many clusters of each class occur in the alloy. A single class j contributes to the macroscopic alloy with its cluster fraction x_j that is defined by $x_j = M_j/M$. The x_j fulfill the constraint

$$\sum_{j=0}^{J} x_j = 1, \qquad (2.83)$$

which stems directly from the relation $M = \sum_j M_j$ for the cluster set. The n cation sites of each cluster are occupied with n_j atoms of species A and $(n - n_j)$ atoms of species B. Since the molar fraction of A atoms for the entire alloy $A_x B_{1-x} C$ is fixed by x, the cluster fractions x_j have to obey the second constraint

$$\sum_{j=0}^{J} n_j x_j = nx. \qquad (2.84)$$

Using such a cluster expansion any macroscopic alloy can be built from the microscopic clusters, each of which contributes with its cluster fraction. Consequently, within this framework each property P of the macroscopic system can be traced back to the respective properties P_j of the clusters. Given the weights $x_j(x,T)$ for an alloy of a certain composition x at a temperature T and the values P_j of the property for each cluster, one can calculate the property $P(x,T)$ for the alloy using the Connolly-Williams method [72, 73],

$$P(x,T) = \sum_{j=0}^{J} x_j(x,T) P_j. \qquad (2.85)$$

For both the rs and the wz crystal structure of the oxides studied in this work we use 16-atom cells for the cluster expansion, relying on previous arguments [74, 75] that it is sufficient to include next-nearest-neighbor correlations in order to capture large parts of the physics. For the wz structure we use the expansion described in Ref. [74], whereas we derived a new expansion for alloys with rs crystal structure (cf. Appendix A.1 and Ref. [76]).

2.5.2 Generalized quasi-chemical approximation

In the framework of the generalized quasi-chemical approximation (GQCA), the cluster fractions are determined by a minimization of the Helmholtz free energy $F(x,T)$ involving all clusters of the expansion [68–71, 73]. To find the cluster fractions x_j^{GQCA} we discuss the mixing contribution ΔF to the free energy

$$\Delta F(x,T) = \Delta U(x,T) - T\Delta S(x,T). \tag{2.86}$$

The mixing contribution to the internal energy $\Delta U(x,T)$ is calculated as the sum of the contributions from the M clusters, referenced to the internal energy U of an alloy consisting only of the two binary end components AC and BC,

$$\Delta U(x,T) = M\left(\sum_{j=0}^{J} x_j \varepsilon_j - [x\varepsilon_J + (1-x)\varepsilon_0]\right) = M\sum_{j=0}^{J} \Delta\varepsilon_j x_j. \tag{2.87}$$

In Eq. (2.87) we introduced the definition of the excess energy $\Delta\varepsilon_j$ for the class j,

$$\Delta\varepsilon_j = \varepsilon_j - \left(\frac{n_j}{n}\varepsilon_J + \frac{n-n_j}{n}\varepsilon_0\right). \tag{2.88}$$

We still need an expression for the configurational (or mixing) entropy in order to calculate the free energy, Eq. (2.86). To evaluate the Boltzmann definition of the entropy,

$$\Delta S(x,T) = k_B \ln W, \tag{2.89}$$

one has to give an expression for the number of possible configurations W. Given a cluster expansion and, therefore, a set of cluster fractions $\{x_j\}$ that fulfills the constraints (2.83) and (2.84), W describes the number of possible atomic configurations in the entire alloy, i.e., W counts all possible ways of arranging the N_A A atoms and N_B B atoms for *one* given set $\{x_j\}$ on the $N = N_A + N_B$ cation sites. To determine W, the number of ways of arranging the $M_0, M_1, \ldots, M_J$ clusters to form the alloy, $M!/\prod_j M_j!$, needs to be multiplied by the number of possibilities of arranging the cations in each cluster. Since one cluster of class j can be occupied by cations in g_j ways, all M_j clusters lead to $g_j^{M_j}$ possibilities. Taking into account all classes j, one ultimately obtains

$$W = \frac{M!}{\prod_{j=0}^{J} M_j!} \cdot \prod_{j'=0}^{J} g_{j'}^{M_{j'}}. \tag{2.90}$$

With Eq. (2.90) and the definition $x_j = M_j/M$, the mixing entropy ΔS in the Stirling limit is

$$\Delta S(x,T) = -k_B M \sum_{j=0}^{J} x_j \ln\left(\frac{x_j}{g_j}\right). \tag{2.91}$$

2.5 Alloy statistics and thermodynamics

Using the *ideal* cluster fractions of a strict-regular solution [71] (see Section 2.5.3),

$$x_j^0 = g_j x^{n_j} (1-x)^{n-n_j}, \qquad (2.92)$$

the mixing entropy can be rewritten as [71, 73]

$$\Delta S(x,T) = -k_\mathrm{B} \left\{ N[x\ln x + (1-x)\ln(1-x)] + M \sum_{j=0}^{J} x_j \ln\left(\frac{x_j}{x_j^0}\right) \right\}. \qquad (2.93)$$

Although the expressions for W according to Eq. (2.90) and that derived in Ref. [73] differ, they both lead to the same entropy in the Stirling limit.

Equations (2.87) and (2.91) determine the Helmholtz mixing free energy as a function of x and T, given that the cluster fractions x_j are known for x and T. In the GQCA the x_j are determined by the requirement that $\Delta F(x,T)$ assumes a minimum with respect to the cluster distribution, i.e., $\partial \Delta F(x,T)/\partial x_j = 0$. Hence, the Lagrange formalism with the constraint (2.83) yields

$$x_j^{\mathrm{GQCA}}(x,T) = \frac{g_j \eta^{n_j} e^{-\beta \Delta \varepsilon_j}}{\sum_{j'=0}^{J} g_{j'} \eta^{n_{j'}} e^{-\beta \Delta \varepsilon_{j'}}}, \qquad (2.94)$$

where $\beta = 1/k_\mathrm{B}T$. The parameter η has to be determined from the condition that the x_j^{GQCA} obey the constraint (2.84). The cluster fractions x_j^{GQCA}, according to Eq. (2.94), describe the probability for the occurrence of each cluster class j in an alloy which has been prepared under thermodynamic equilibrium conditions that minimize the free energy.

2.5.3 Strict-regular solution and microscopic decomposition limit

Besides the thermodynamic equilibrium described above, the experimental situation also suggests the studying of certain non-equilibrium preparation conditions, for which the actual cluster statistics may be modified by kinetic barriers, frozen high-temperature states, as well as interface or surface influences. In order to simulate a dependence of the cluster distribution on the preparation conditions we study two limiting cases:

(i) The *strict-regular solution (SRS) model* [71]: In this case, the ideal cluster fractions x_j^0 [cf. Eq. (2.92)] are used, which arise from a purely stochastic distribution of the clusters. These x_j^0 do not depend on the temperature or on the clusters' excess energies but are only determined by x and n_j. The number W in this case is simply given by all possible arrangements of $N_\mathrm{A} = xN$ A atoms and $N_\mathrm{B} = (1-x)N$ B atoms on the $N = N_\mathrm{A} + N_\mathrm{B}$ cation sites of the alloy, i.e.,

$$W^{\mathrm{SRS}} = \frac{N!}{N_\mathrm{A}! N_\mathrm{B}!}. \qquad (2.95)$$

Using Eq. (2.89) we obtain for the mixing entropy in the Stirling limit

$$\Delta S^{\text{SRS}}(x) = -k_{\text{B}} N [x \ln x + (1-x) \ln(1-x)]. \quad (2.96)$$

The ideal x_j^0, according to Eq. (2.92), and ΔS^{SRS} can be interpreted as the high-temperature limit of the GQCA since the x_j^{GQCA} approach the x_j^0 as the temperature increases.

(ii) The *microscopic decomposition model (MDM)*: In this limiting case the cations of type A (B) are more likely to occur close to cations of type A (B). Consequently, only the clusters representing the binary components AC and BC are allowed, with xM being the number of AC clusters and $(1-x)M$ the number of BC clusters. This is equivalent to a linear interpolation between the binary end components. For positive excess energies $\Delta \varepsilon_j$ [cf. Eq. (2.88)] the x_j^{MDM} represent the low-temperature limit of the GQCA. The cluster fractions for the MDM are given by

$$x_j^{\text{MDM}} = \begin{cases} 1-x & \text{for } j=0 \\ x & \text{for } j=J \\ 0 & \text{otherwise} \end{cases} \quad (2.97)$$

The number of atomic configurations follows immediately from Eq. (2.90) by taking only the two clusters $j=0$ and $j=J$ into account, i.e.,

$$W^{\text{MDM}} = \frac{M!}{[xM]! [(1-x)M]!}. \quad (2.98)$$

Using the x_j^{MDM} we obtain for the mixing entropy [cf. Eq. (2.89)]

$$\Delta S^{\text{MDM}}(x) = -k_{\text{B}} M [x \ln x + (1-x) \ln(1-x)] = \frac{1}{n} \Delta S^{\text{SRS}}(x), \quad (2.99)$$

i.e., a reduced configurational entropy. In general, the MDM describes alloys that have been prepared under conditions where mixing does not lead to a gain of internal energy. Indeed, in the MDM description one finds $\Delta U(x) = 0$ for the mixing contribution to the internal energy [cf. Eq. (2.87)].

3 Practical issues

> Wer gar zuviel bedenkt,
> wird wenig leisten.
>
> Johann Christoph Friedrich von Schiller
> Wilhelm Tell III, 1

After introducing the theoretical fundament of this work in the preceding chapter, the aspects that must be considered for the actual calculations are elucidated in the following. Here, we focus on the more technical details, whereas the computational parameters that we used for our calculations are summarized in Appendix A.2.

Most of the calculations are done using version 5.1.39 of the *Vienna Ab-Initio Simulation Package* (VASP) [77, 78]. This is a software package for solving the KS equations as well as the generalized KS equations in reciprocal space. The wave functions are expanded into plane waves. To model the electron-ion interaction the projector-augmented wave (PAW) method is applied [42, 79, 80], allowing a highly accurate description of the wave functions with almost the same quality as produced by all-electron calculations. Furthermore, the VASP code features the computation of QP energies within the GW approximation, based on the PAW method and using the fully frequency-dependent DF [48, 81–83].

To obtain the optical properties, taking excitonic and local-field effects into account via the solution of the BSE, we employ an implementation that has been continuously developed in our group and is based on version 4.4 of VASP [50, 52, 57, 61–63, 84–86]. The input which is required for calculating the excitonic Hamiltonian, such as wave functions, QP energies, and optical-transition matrix elements, stems from VASP 5.1.39. After calculating the excitonic Hamiltonian, either its lowest eigenvalues and eigenstates are obtained using an iterative-diagonalization scheme [61] or the DF is calculated by solving an initial-value problem [62, 63].

3.1 Electronic properties

3.1.1 Hybrid functional and quasiparticle corrections

When we introduced the G_0W_0 approach in Section 2.3 we briefly mentioned that an appropriate starting electronic structure is an important key point for the computation of the QP corrections by means of the first-order of the perturbation theory. For the transition-metal oxides, several nitrides, as well as the group-II oxides studied in this work, it turned out that the GGA as an approximation to XC shows severe deficiencies in the description of the electronic structure [19, 47, 57, 87]. Not only the fundamental band gaps are strongly underestimated, but the energetic positions of the Zn $3d$ and Cd $4d$ electrons deviate significantly from experimental results [19, 84]. Partly, these problems can be traced back to the neglect of the excitation aspect when comparing KS eigenvalues to experimental results. In addition, as pointed out in Section 2.2.4, the self-interaction is not properly removed by the GGA which, especially for the shallow Zn $3d$ and Cd $4d$ electrons, further corrupts the results within GGA. Hence, the d states appear about 2 eV too close to the O $2p$ states at the VB maximum (VBM), which leads to an overestimation of the hybridization of these p and d levels. Both the strongly underestimated gap of wz-ZnO and the negative gap of rs-CdO obtained within GGA we have already attributed to the overestimation of the pd repulsion [19, 47, 88].

As a consequence, using GGA as an approximation for XC is not sufficient because the corresponding KS eigenvalues are too far from the QP results and QP corrections calculated from first-order perturbation theory cannot cure these deviations. For that reason our *ab-initio* description of the electronic structure of the group-II oxides is based on the HSE functional [38] (cf. Section 2.2.4). In this case, the self-energy in the generalized KS equation is already a good approximation of that in the QP equation (2.47) since the eigenvalues computed using the HSE functional are closer to the final QP-corrected results. Furthermore, the HSE wave functions for the d states are more localized than those obtained within the GGA [47, 50]. We expect the HSE functional to provide meaningful input for the calculation of QP energies within the G_0W_0 approximation, which is confirmed by a reduction of the QP shifts [47, 50] compared to the ones computed with a GGA starting point. We refer to the calculation of QP energies within the G_0W_0 approximation using a HSE starting electronic structure as the HSE+G_0W_0 method. From test calculations we expect the corresponding QP energies to be converged within about 0.1 eV.

3.1.2 Mapping to an affordable approach

For MgO, ZnO, and CdO, results obtained using the HSE+G_0W_0 approach are presented along with experimental findings in Section 4.1. However, due to the extremely high computational cost of this method we are facing a problem when it shall be extended towards the treatment of significantly more **k** points or bands. Especially when calculating the starting electronic structure for the BSE calculation, the high demands for memory and processor power render the application of the HSE+G_0W_0 method not applicable [38]. The main reasons are (i) the computationally expensive, non-local HF exchange contribution, which is part of the HSE functional, and (ii) the large number of empty CBs necessary for computing the DF which enters the screened Coulomb potential within the G_0W_0 approximation of the self-energy.

To circumvent this problem when the input (wave functions, eigenvalues, optical transition-matrix elements) for the calculation of the BSE Hamiltonian is computed, we pursue a different approach by simulating the results of the HSE+G_0W_0 calculations via a GGA+U+Δ method [57, 84, 85, 87, 89]. Here, U denotes the additional Coulomb interaction term within the GGA+U approach (cf. Section 2.2.4) and Δ describes a scissors operator [46] that rigidly shifts all CBs. These two parameters, U and Δ, are determined using the HSE+G_0W_0 results: While U is adjusted in such a way that the energetical position of the d bands obtained from the GGA+U calculation matches the HSE result, we use Δ to enlarge the fundamental gap until it is identical to the HSE+G_0W_0 value. This mapping of the *eigenvalues* is justified for the three oxides in Section 4.1.1. In addition, previous studies reported comparably small differences for the *wave functions* that resulted from HSE calculations or the GGA+U approach [47, 50], or found a high overlap of QP wave functions and KS-LDA ones [43]. As discussed in Refs. [50, 57] a strict criterion for comparing wave functions does not exist and their actual suitability for a certain type of calculations is difficult to evaluate. However, the prohibitively high computational cost of HSE+G_0W_0 calculations forces us to use an approximation, such as the GGA+U+Δ method, when calculating the starting electronic structure to solve the BSE.

3.1.3 Inclusion of spin-orbit coupling

The high computational cost of HSE+G_0W_0 calculations would further increase when accounting for the full spinors. For the oxides studied in this work, the influence of the spin-orbit-interaction on the KS energies around the fundamental gap is small (cf. Section 4.1.2) and, hence, we expect only a small impact on the optical transition-matrix elements that enter, via ε, the screened Coulomb potential W. Therefore, to obtain QP energies including the SOC-related effects, we employ an approach that is inspired by perturbation theory. We assume that the influence of the spin-orbit interaction on the QP corrections is negligible which is reasonable at least for absolute spin-orbit induced shifts that are smaller than the QP corrections. Hence, we apply the QP shifts, as calculated for spin-paired electrons, to HSE eigenvalues that have been obtained from a calculation where the SOC has been taken into account [42, 90, 91].

3.2 Optical properties

As elucidated in Section 2.4 the BSE for the polarization function must be solved to calculate the macroscopic DF that includes excitonic and local-field effects. The corresponding excitonic Hamiltonian is computed from an initial electronic structure that is obtained using the GGA+U+Δ approach (cf. Section 3.1.2). The optical transition-matrix elements that enter into Eqs. (2.64) and (2.67) are calculated in the longitudinal approximation [83]. To account for any broadening mechanisms that are relevant in the experiment, e.g. those due to finite lifetimes or temperatures, a Lorentzian broadening of 0.1 eV is applied to the calculated DFs.

For actual calculations of converged results for optical spectra and exciton binding energies we have to fulfill two partially competing requirements: While the investigation of high-energy transitions necessarily demands for the inclusion of a large number of CBs, a thorough description of the low-energy optical transitions requires a fine **k**-point sampling of the BZ to converge Wannier-Mott-like bound excitonic states [61]. Both requirements independently lead to very high computational costs, not only for the calculation of the starting electronic structure, but also of the excitonic Hamiltonian. In addition, storing the resulting matrices requires from 10 to several 100 gigabytes worth of memory and hard-disk space.

Therefore, we employ a technique for an adapted sampling of the BZ [61] to deal with both issues (cf. Sec. 3.2.1). In addition, spin-orbit coupling is included via a perturbative approach [84] (cf. Sec. 3.2.2). Moreover, below we give details about the model function [92, 93] that we use to describe the screening of the electron-hole interaction.

3.2.1 Adapted sampling of the Brillouin zone

For the materials studied in this work, the lowest optical transitions can be traced back to a relatively small **k**-space region around the Γ point [19]. Hence, this part of the BZ must be sampled very densely in order to obtain converged results for the optical properties in the vicinity of the absorption edge [61]. Increasing the sampling density for the entire BZ leads to unreasonably large amounts of **k** points. Therefore, we employ hybrid **k**-point meshes [61] which feature a coarse sampling of the outer regions of the BZ and **k** points concentrated

3.2 Optical properties

in the vicinity of Γ. These meshes are described by three values, $X:Y:Z$ (see Ref. [61]), indicating that the entire BZ is covered by a regular, coarse $X \times X \times X$ Monkhorst-Pack (MP) mesh [94] which inner $Y \times Y \times Y$ boxes are sampled by $Z \cdot Y/X + 1$ MP points along each direction, respectively. This leads to a sampling density in the inner region that is equal to a $Z \times Z \times Z$ MP mesh applied to the entire BZ. In addition, we shift all **k** meshes that are used for the calculation of the DFs by a small random vector that lies within one box of the coarse MP mesh. This shift of the entire mesh lifts symmetry degeneracies inherently present in MP **k**-point sets and, therefore, improves the convergence of the respective optical quantities [89].

This technique enables us to obtain the optical properties close to the absorption edge, though, for computational reasons, it is not possible to simultaneously increase the number of CBs to study high-energy optical transitions. Therefore, we solve two separate BSEs; one for the low energy range close to the absorption edge, using the dense, hybrid **k**-point meshes and one extending to high-energy transitions at a reduced BZ sampling density. The resulting DFs are merged [89], which is straightforward for the imaginary parts. While the omission of the high-energy transitions influences only the high-energy region of the imaginary part, it affects the *entire* real part. In particular, it causes an underestimation of the electronic static dielectric constant $\varepsilon_\infty = \text{Re } \varepsilon(\omega = 0)$ due to the Kramers-Kronig relation [95, 96] between the real and imaginary parts. Therefore, the real part of the low-energy DF is shifted so that its dielectric constant matches that of the real part belonging to the high-energy DF. For the same reason we extend the DF above photon energies of 32.5 eV by adding the contributions that correspond to transition energies higher than 32.5 eV as calculated within the independent-QP approximation (IQPA), i.e., we take QP energies into account while neglecting the electron-hole interaction in this energy range (see Section 4.2.1). This slightly improves the real part at low energies and is well-justified since in the high-energy range the absolute value of the imaginary part is already below 10 % of the peak values and excitonic effects can then safely be neglected.

3.2.2 Inclusion of spin-orbit coupling

To include the influence of the spin-orbit interaction on the QP energies that enter the computation of the optical properties [cf. Eqs. (2.62), (2.64), and (2.67)], we follow an approach that is akin to the one presented in Section 3.1.3. It allows us to avoid the full spinors again and it is justified *a posteriori* by the smallness of the spin-orbit effects. Here, we calculate **k**-dependent spin-orbit shifts as the difference between a band structure including SOC and one calculated without the spin-orbit interaction. Both are obtained by means of the

GGA+U scheme employing the same **k** mesh as used for setting up the excitonic Hamiltonian. These shifts are then added, for both spin components separately, to the eigenvalues of the GGA+U+Δ calculation used for setting up the excitonic Hamiltonian and two separate BSEs are solved [84].

3.2.3 Screening of the electron-hole interaction

The calculation of the screened electron-hole interaction requires a reasonable approximation for the screening in the system. We go beyond merely using a static dielectric constant ε_{eff} [e.g. Eq. (2.68)] by employing a model function [92, 93] which is given by

$$\varepsilon(|\mathbf{q}|,n_v) = 1 + \left[\frac{1}{\varepsilon_{\text{eff}}-1} + \alpha\left(\frac{q}{q_{\text{TF},n_v}}\right)^2 + \frac{3q^4}{4k_{\text{F},n_v}^2 q_{\text{TF},n_v}^2}\right]^{-1}, \qquad (3.1)$$

with the parameter $\alpha = 1$. Equation (3.1) is an interpolation between ε_{eff} of the semiconductor for $q=0$, a TF-like behavior for small q, and a free-electron gas behavior for large q. For ε_{eff} in Eq. (3.1) we use the values that we obtained within the independent-particle approximation (see Appendix A.2). To obtain the Fermi and the TF wave vectors within the parameter-free approach, they are calculated using the cell-averaged density n_v of the valence electrons (including Zn $3d$ and Cd $4d$ electrons).

4 Ideal MgO, ZnO, and CdO

> Ideals are like stars; you will not succeed in touching them with your hands. But like the seafaring man on the desert of waters, you choose them as your guides, and following them you will reach your destiny.
>
> Carl Schurz

For modern consumer electronics, solar cells, or intelligent materials applications simultaneous interplay of transparency in the optical spectral range and electrical conductivity under ambient conditions is undoubtedly of large interest. An efficient combination of electrical and optical components is also desirable to exponentiate the signal processing speed in, e.g., networking technology. Nowadays, the research and development in this context is oftentimes associated with the term "transparent electronics" or, occasionally, "oxide electronics". A large fundamental band gap renders the group-II oxides transparent in the visible spectral range, which is a very important property of these materials. Additionally, it has been reported that, to some extent, the conductivity of ZnO can be achieved via n-doping [97–99]. Due to their larger (e.g. MgO) or smaller (e.g. CdO) fundamental band gap other group-II oxides are discussed as candidates for combinations with ZnO in the form of alloys or heterostructures — to achieve the goal of tailoring different properties of materials [100–104].

Furthermore, it is certainly beneficial that for ZnO as well as for MgO very pure, high-quality single crystals are commercially available. ZnO is especially easily available as a resource, in addition to being environmentally friendly and biocompatible. Interestingly, a large variety of nanostructures have been observed in several experiments; among them are nano-rods, -rings, -brushes, and -tubes [105–107].

Thus, such fascinating properties are the reason for an ongoing boost of interest in group-II oxides. A large number of possible applications explains the enormous technical importance of these materials and, consequently, one finds more than $1,000$ publications per year since 1997 that contain "ZnO" or "zinc oxide" in their title and over $5,000$ such articles in 2008, as well as 2009 [108]. Furthermore, the oxides are well-suited for studying very fundamental physics, explaining why we chose to focus on MgO, ZnO, and CdO in this work.

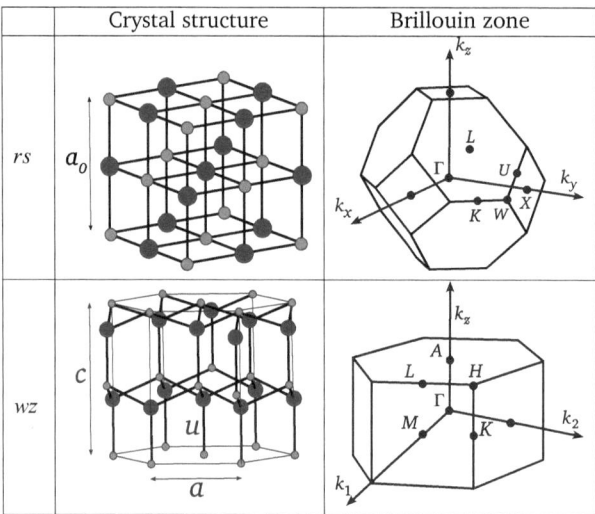

Table 4.1: Real-space structure and Brillouin zone of the hexagonal *wz* and the cubic *rs* lattice. The parameters that describe the lattice as well as several high-symmetry points in the Brillouin zone are denoted in the figure.

In this chapter we investigate electronic and optical properties for the equilibrium crystal structures (under ambient conditions), i.e. *rs*-MgO, *wz*-ZnO, and *rs*-CdO. Both the *wz* and the *rs* geometry are visualized along with the corresponding Brillouin zones (BZs) in Table 4.1. Here we employ lattice parameters that we derived from total-energy minimizations before [19, 23] and which are summarized in Appendix A.2. Studying the bulk materials provides necessary and helpful insight into the electronic structure (Section 4.1) as well as the optical properties (Section 4.2). This knowledge will form the basis for our investigations regarding the influence of imperfections in the following chapters. Detailed comparison to experimental results are made throughout and, in addition, we use our *ab-initio* results for the band structures and the DF to derive information about the branch-point energies in Section 4.1.3 and the electron-energy loss function in Section 4.2.4.

4.1 One-particle excitations

4.1.1 Band structures and densities of states

Band structures and densities of states (DOS) are both experimentally accessible quantities describing the electronic structure of a material. A goal of this work is their highly accurate computation within the HSE+G_0W_0 approach. Our calculated curves are shown in Fig. 4.1

4.1 One-particle excitations

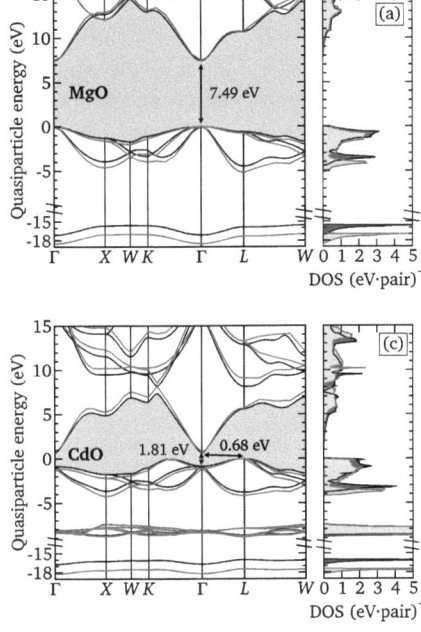

Figure 4.1: Quasiparticle band structures and the densities of states of *rs*-MgO (a), *wz*-ZnO (b), and *rs*-CdO (c) as calculated within the HSE+G_0W_0 approach (gray lines) and the GGA+U+Δ approach (black lines). The valence band maximum is used as energy zero and the fundamental gap region is shaded. Direct and indirect gaps are given in the figures.

where we plot the DOS alongside the QP band structure for the three oxides. Since we also aim for calculating the optical properties later, the HSE+G_0W_0 findings are compared to results calculated within the computationally less demanding GGA+U+Δ approximation (cf. Chapter 3) in the same figure.

Discussion of the HSE+G_0W_0 results

For *rs*-MgO we find a fundamental direct band gap of 7.49 eV at the Γ point of the BZ [see Fig. 4.1(a)]. It is formed by O 2*p* states that extend, being the uppermost valence states, from the VBM to about −5 eV below and Mg 3*s* states which represent the lowest CB. For *wz*-ZnO the fundamental gap is also found at the Γ point of the BZ and amounts to 3.21 eV. It separates Zn 4*s* CB states from the uppermost *pd*-hybridized VB states that show 75 % O 2*p* character and 25 % Zn 3*d* contributions. As seen in Fig. 4.1(b), this upper VB complex extends to about −5 eV. In addition, we find ten occupied *d* bands at roughly 6 to 8 eV below the VBM. Such a complex is also observed for *rs*-CdO, where the five Cd 4*d* bands appear at −7 to −9 eV, with respect to the maximum of the VBs. Following the chemical trend that occurs along the row MgO, ZnO, CdO, the direct gap at the Γ point in this material, formed by O $2p$ valence states without any d character and a Cd $4s$-derived CB, is with 1.81 eV somewhat smaller. In addition, the uppermost VB at the L point is 1.12 eV above the highest valence state at the BZ center [see Fig. 4.1(c)] and the lowest gap turns out to be an indirect one of about 0.68 eV. This occurs because in the rs crystal structure, contrary to the wz case, the hybridization of p and d states at the Γ point is forbidden for symmetry reasons [19, 88]. This symmetry constraint does not apply apart from the BZ center, leading to an increase in energy of the corresponding states due to the pd hybridization. For all three oxides the weakly dispersive O $2s$ states occur at roughly -18 eV.

Comparison to experimental results: Band structures

Our calculated values for the fundamental energy gaps, $E_g = 7.49$ eV, 3.21 eV, and 0.68 eV for rs-MgO, wz-ZnO, and rs-CdO, respectively, only slightly underestimate measured results of 7.67 eV [109], 3.4 eV [110], and 0.84 eV [111]. Comparing the value of 1.81 eV for the direct gap at Γ of rs-CdO with an experimental result of 2.28 eV [111] also shows an underestimation.

We attribute this trend of slightly smaller fundamental gaps to two different causes. The first of these is related to the DFT-GGA results for the atomic geometries that we used. In Ref. [19] they have been shown to slightly exceed measured lattice constants which, in turn, leads to too small gaps due to the reduced confinement of the electrons. Such an influence on the gap can amount to about 0.1 eV per 2 % lattice mismatch, as estimated from uniaxially strained wz-ZnO (see Section 5.1), but also depends on the corresponding deformation potentials for each individual material. A second problem arises from the calculation of QP corrections using first-order perturbation theory. In the HSE band structure the gaps are still up to 1.6 eV (rs-MgO) smaller than the experimental values. Also, the HSE description slightly underestimates the d-band binding energies with respect to the VBM in comparison to experimental results [84, 112–114] for wz-ZnO and rs-CdO. That leads to an additional closing of the gap due to the pd repulsion as discussed earlier in Section 3.1. Therefore, we find that even though the HSE functional produces a much better starting point for the electronic-structure calculations than the DFT-GGA does, its results still constitute a challenge to the one-step perturbation-theory approach to the calculation of the QP energies.

M. Kobayashi et al. experimentally determined the upper VB structure for wz-ZnO by means of soft X-ray angle-resolved photoemission spectroscopy [115]. Figure 4.2 shows the results of this measurement together with QP energies that we calculated using the HSE+G_0W_0 approach. The agreement of energetic positions at high-symmetry points in the BZ, and

4.1 One-particle excitations

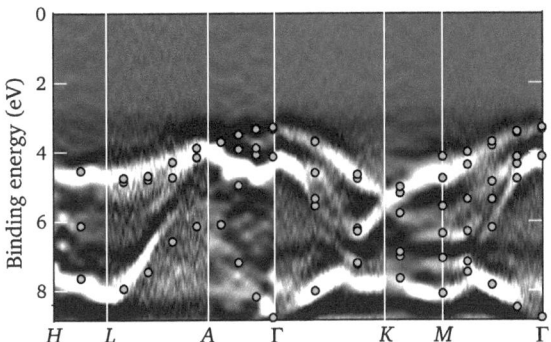

Figure 4.2: Comparison of the HSE+G_0W_0 QP energies (dots) with a band structure that has been derived from soft X-ray angle-resolved photoemission spectroscopy measurements for wz-ZnO [115]. High intensities in the measurement (white) indicate bands.

also along the high-symmetry lines in-between, is satisfying. Our approach reproduces the measurement impressively well, especially along the path $\Gamma-K-M-\Gamma$. At the A point and its surroundings slight deviations occur mainly because the lowest plotted band between A and Γ is almost invisible in the experiment, an effect which we attribute to matrix-element contributions. While their influence is not taken into account in our calculations, they might be the reason for the low intensities observed in the experiment in this case since the QP energies itself agree well with the low-intensity structure observed in this experiment.

Comparison to experimental results: Densities of states

For rs-MgO and rs-CdO there is experimental data from soft X-ray emission spectroscopy available [84], an experimental technique that is element-specific and directly accesses the orbital-angular-momentum-resolved partial DOS (PDOS) [84, 116]. In Figs. 4.3(a) and (c) we compare the measured data to the O $2p$ PDOS, calculated by means of the HSE+G_0W_0 approach. In addition, in Fig. 4.3(b) we plot the total DOS computed for wz-ZnO to compare to data from high-resolution X-ray photoemission spectroscopy (XPS) [84]. We incorporated a Gaussian broadening of 0.45 eV full width at half maximum into the theoretical DOS curves in Fig. 4.3 to take the effect of a certain instrumental resolution into account.

In the case of rs-MgO [cf. Fig. 4.3(a)] the measurement agrees with our calculated results for both the O $2p$ band width as well as the two distinct peaks at -4.2 eV and -1.2 eV below the VBM that arise due to the uppermost VB complex. Also for rs-CdO we find very good agreement between calculated and experimental results regarding the positions of the peaks at -3.6 eV and -1.1 eV below the VBM as well as for the band width [see Fig. 4.3(c)]. For both materials, rs-MgO and rs-CdO, the heights of the peaks at -4.2 eV and -3.6 eV are overestimated relative to the height of the second peak. We attribute this to matrix-element

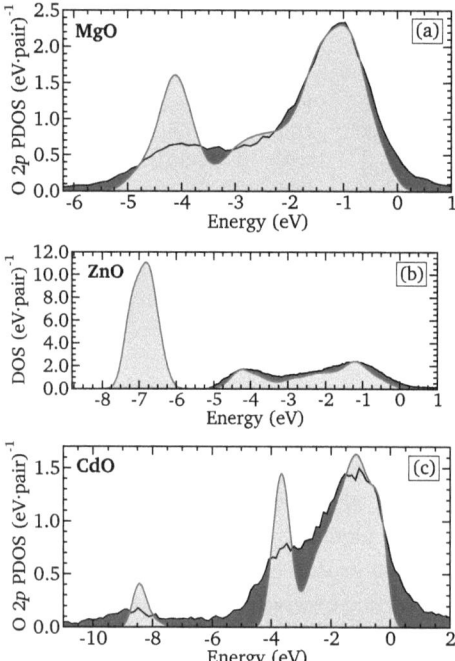

Figure 4.3:
(a) O $2p$ partial density of states of rs-MgO obtained from soft X-ray emission measurements [84] (black) and from the HSE+G_0W_0 approach (gray).
(b) Density of states of wz-ZnO from an X-ray photoemission spectroscopy experiment [84] (black) and the HSE+G_0W_0 approach (gray).
(c) O $2p$ partial density of states of rs-CdO obtained from soft X-ray emission measurements [84] (black) and from the HSE+G_0W_0 approach (gray).
The valence-band maximum has been used as energy zero in all cases.

effects that influence the transitions but are neglected in our calculations. In the case of rs-CdO, the measurement confirms an additional spectral feature which is traced back to the aforementioned hybridization of the O $2p$ and the Cd $4d$ states. It appears at an energy of about -8.4 eV in the calculated PDOS, whereas in the measured curve this peak occurs about 0.4 eV lower in energy. Comparison to an X-ray absorption spectroscopy experiment [113] shows good overall agreement for the O $2p$ PDOS close to the CBM, though the relative intensities do differ.

In Fig. 4.3(b) we compare a curve from an XPS measurement [84] to the computed total DOS of the O $2p$-related uppermost VB complex for wz-ZnO and find a good agreement regarding the heights and positions of the two pronounced maxima. In our calculated curve the peak caused by the Zn $3d$ states occurs at about -6.8 eV (with respect to the VBM) and, therefore, is slightly higher in energy than the -6.95 eV obtained in an angle-resolved photoelectron spectroscopy experiment [117]. In another soft X-ray emission measurement [112] this peak is found at about -7.5 eV below the VBM.

4.1 One-particle excitations

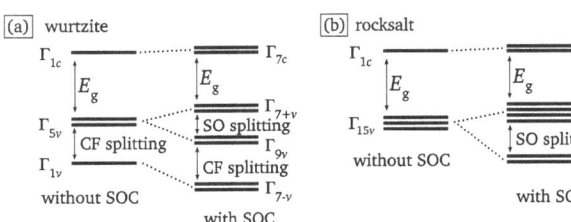

Figure 4.4: Classification of the lowest conduction- and uppermost valence-band states at the Γ point of the Brillouin zone for the (a) *wz* and (b) *rs* crystal structure.

Mapping to the GGA+U+Δ approximation

In the following we discuss the approximation of the HSE+G_0W_0 results by means of the GGA+U+Δ approach, using the two parameters U and Δ (cf. Chapter 3). For the gap calculated within GGA for *rs*-MgO we find an underestimation of $\Delta = 2.99$ eV. For *wz*-ZnO and *rs*-CdO the remaining gap differences between the GGA+U and HSE+G_0W_0 values are $\Delta = 1.78$ eV and $\Delta = 1.07$ eV, respectively. The effective Coulomb repulsion U of the GGA+U method [25], used to correct the energetic position of the strongly localized cationic semi-core $3d$ ($4d$) electrons with respect to GGA results, has been set to $U = 6.5$ eV ($U = 4.5$ eV) for *wz*-ZnO (*rs*-CdO). As seen in Fig. 4.1, these choices for U and Δ allow us to obtain good agreement for the band structures and the DOS. The GGA+U+Δ scheme is well-suited for generating the starting electronic structure for the optical calculations covering a wide range of photon energies [89].

Nevertheless, such a mapping has certain deficiencies. For all three oxides, the band width of the uppermost O$\,2p$-derived VB complex, obtained by the GGA+U+Δ approach, is underestimated by about 0.5 eV with respect to the HSE+G_0W_0 results. Of course, this effect will slightly affect the optical properties in the low-energy range. Also, for all three materials, the binding energy of the O$\,2s$ electrons is underestimated by about 1.4 eV within the GGA+U+Δ approach, as can be seen in the band structure and the DOS in Fig. 4.1 around -18 eV. A possible influence on the optical properties in the high-photon-energy region above $\hbar\omega = (18.0 + E_g)$ eV remains small for all three oxides because of the low oscillator strengths of the corresponding high-energy optical transitions. More importantly, Fig. 4.1 reminds us that the scissors operator Δ does not reproduce any of the energy dependence of the self-energy operator Σ. This explains a general trend that is found in Fig. 4.1; the higher-lying CBs calculated via the GGA+U+Δ method are slightly too low in energy.

4.1.2 Inclusion of spin-orbit coupling

Quasiparticle band structure

Taking the spin-orbit interaction, Eq. (2.40), into account causes an influence on the absolute values of the QP energies. In addition, the symmetry of the Hamiltonian of the problem is lowered and, therefore, symmetry-induced degeneracies can be lifted. Hence, in Fig. 4.4 we classify the uppermost VB states and the lowest CB state at the Γ point using the irreducible representations from the character tables of the symmetry groups C_{6v}^4 (O_h^5) for the *wz* (*rs*) crystal structure. Taking the SOC into account leads to the double-group notation [118] and doubles the number of states when going from double-occupied (due to the spin degeneracy) to single-occupied bands. Figure 4.4 shows a non-degenerate CBM of Γ_{1c} type for *wz* polymorphs in the absence of SOC. The uppermost VB states consist of a twofold degenerate Γ_{5v} level which is separated from a Γ_{1v} level due to the crystal-field (CF) in the hexagonal crystal system. Inclusion of SOC transforms the Γ_{1v} state into two degenerate Γ_{7-v} levels that contain a large contribution of the atomic p_z orbital, which is oriented parallel to the hexagonal *c* axis. The Γ_{5v} state splits into Γ_{9v}/Γ_{7+v} with no/small p_z contributions [118, 119]. In the case of the *rs* crystal structure [cf. Fig. 4.4(b)] the uppermost threefold degenerate Γ_{15v} VB level splits into a fourfold degenerate Γ_{8v}^- and a twofold degenerate Γ_{6v}^- state, whereas the lowest CB state Γ_{1c} (without SOC) becomes a twofold degenerate Γ_{6c}^+ state.

Our calculations indicate that the spin-orbit-induced splittings of the uppermost VB levels of the three oxides are small compared to the fundamental gaps. From the projections of the wave function onto atomic *s*, *p*, or *d* states we obtain the character of the wave function and are able to assign the irreducible representations discussed earlier. In the case of *wz*-ZnO we find that the uppermost valence states are Γ_{7+}-derived, followed by two Γ_9-derived states, which lie 11.3 meV lower in energy. The CF split-off bands are of Γ_{7-} type and occur 48.3 meV below the Γ_9 states (cf. Table 4.2).

Also, for *rs*-MgO and *rs*-CdO, the splittings between Γ_{8v}^- and Γ_{6v}^- remain small at 43.6 meV and 82.0 meV, respectively. Comparison to the value of 58.3 meV calculated for *rs*-ZnO [91] confirms the expected chemical trend of an increasing influence of the spin-orbit interaction on the splittings with increasing mass of the cation. Nevertheless, we find the influence of the spin-orbit interaction at the VBM of *wz*-ZnO to be smaller than for the two oxides in the *rs* crystal structure. This can be traced back to the aforementioned hybridization of atomic Zn 3*d* and O 2*p* states which is symmetry-forbidden at Γ in the *rs* crystal structure but occurs in *wz* crystals. Depending on the sign and order of magnitude of the respective coefficients that mix O 2*p* and Zn 3*d* wave functions, the sign and order of magnitude of the resulting splittings of the uppermost VB levels can vary [120] and deviate from the atomic spin-orbit

4.1 One-particle excitations

without SOC	$E_g = \varepsilon^{QP}(\Gamma_{1c}) - \varepsilon^{QP}(\Gamma_{5v})$	3.21
	$\Delta_1^{\text{no SOC}} = \varepsilon^{QP}(\Gamma_{5v}) - \varepsilon^{QP}(\Gamma_{1v})$	54.0
with SOC	$\varepsilon^{QP}(\Gamma_{9v}) - \varepsilon^{QP}(\Gamma_{7+v})$	−11.3
	$\varepsilon^{QP}(\Gamma_{9v}) - \varepsilon^{QP}(\Gamma_{7-v})$	48.3
	Δ_1^{qc}	52.6
	$\Delta_2^{qc} = \Delta_3^{qc}$	−5.2

Table 4.2: Band parameters around the fundamental gap of wz-ZnO: Gap E_g (in eV), valence-band splittings $\varepsilon^{QP}(\Gamma_{9v}) - \varepsilon^{QP}(\Gamma_{7+v})$, and $\varepsilon^{QP}(\Gamma_{9v}) - \varepsilon^{QP}(\Gamma_{7-v})$ (in meV) as well as the derived quantities $\Delta_1, \Delta_2, \Delta_3$ (in meV).

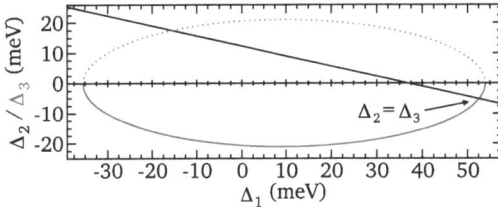

Figure 4.5: Spin-orbit-splitting constants Δ_2 (black) and Δ_3 (gray) as a function of the Δ_1 constant that is related to the crystal-field splitting. All quantities are given in meV and the quasi-cubic approximation for positive Δ_1 is indicated by the arrow.

splittings [121].

For further comparison we analyze our results for wz-ZnO using the expression

$$\varepsilon^{QP}(\Gamma_{9v}) - \varepsilon^{QP}(\Gamma_{7+/-v}) = \frac{1}{2}\left[(\Delta_1 + 3\Delta_2) \mp \sqrt{(\Delta_1 - \Delta_2)^2 + 8\Delta_3^2}\right] \quad (4.1)$$

from **k**·**p** theory [119]. By means of Eq. (4.1) and with the energy differences of the valence levels in a wz crystal we can calculate the two spin-orbit-splitting parameters Δ_2 and Δ_3 as well as the Δ_1 parameter, which is related to the crystal-field splitting. Since, however, the system is underdetermined, we employ the quasi-cubic approximation by assuming $\Delta_2 \equiv \Delta_3$ which then allows the calculation of the values in Table 4.2. From these results it becomes clear that the absolute values of the spin-orbit related constants $\Delta_2^{qc} = \Delta_3^{qc}$ amount to only about 10 % of the one for the CF splitting (Δ_1^{qc}). Alternatively, we compare the trends for Δ_2 and Δ_3, along with the anisotropy between the two, as a function of Δ_1 in Fig. 4.5. From this plot we find that the absolute value of Δ_3 is always smaller than ≈ 20 meV, whereas its sign is undetermined [cf. Eq. (4.1)]. The plot also shows that Δ_2 can change its sign when Δ_1 is only approximately 70 % of the value obtained within the quasi-cubic approximation. However, we expect that the influence of the spin-orbit interaction on the CF splitting is not too large and, in particular, does not change the sign of Δ_1. Taking this into account justifies the signs of the values in Table 4.2.

Comparison to measured spin-orbit splittings

Several detailed experimental investigations have been conducted on the band structure of wz-ZnO. Comparing our results for Δ_1 (cf. Table 4.2) to measured values shows that 38.3 meV from two- and three-photon spectroscopy [122], 39.4 meV from two-photon absorption [123], or 30.5 meV [110, 124] determined from reflectance spectra, are slightly smaller. Another detailed theoretical study, based on DFT-LDA, reports $\Delta_1 = 38$ meV [125]. Our results for the spin-orbit-related Δ_2 and Δ_3 constants are slightly larger than measured values of $\Delta_2 = \Delta_3 = -3.5$ meV [123] (also within the quasicubic approximation). In contrast, $\Delta_2 = -2.1$ meV / $\Delta_3 = -9.1$ meV [122] agree well with our findings shown in Fig. 4.5, while e.g. the ordering of the computed values of $\Delta_2 = -4.5$ meV / $\Delta_3 = -3.1$ meV given in Ref. [125] deviates.

We want to point out that, since the experimental results originate from *optical* measurements, there is an additional influence of the electron-hole interaction on these two splittings. This issue is discussed in more detail in Section 4.2.3. Here we will focus on the sign of these constants and, therefore, the ordering of the uppermost VBs because this has been debated since the 1960's [126]. At present, the consensus seems to confirm the VB ordering found in this work, i.e., $\Gamma_7-\Gamma_9-\Gamma_7$ [127]. Among the recent measurements that indicate a $\Gamma_9-\Gamma_7-\Gamma_7$ ordering [110, 124, 128], at least the findings by D. C. Reynolds *et al.* [128] can be traced back [125] to their wrong assumption regarding the sign of the g factors of the Γ_7 VB.

Unfortunately, to our knowledge there are no measured spin-orbit splittings for the two rs compounds MgO and CdO. In their theoretical work Zhu *et al.* [129] reported values from DFT-LDA calculations yielding 38 meV (68 meV) for rs-MgO (rs-CdO), as well as 51 meV for rs-ZnO. This is in good agreement with our findings for the rs polymorphs of the three oxides. The fact that our numbers are larger by 5.6 meV (MgO), 7.3 meV (ZnO), and 14 meV (CdO) can be attributed to the inclusion of the QP corrections which render our results more reliable.

Effective masses

Via parabolic fitting to the QP band structures (including SOC) in the direct vicinity of the Γ point we derived the diagonal components of the effective mass tensor. The values for three different directions in **k** space are compiled in Table 4.3 for the lowest CB and the three uppermost VBs. We want to remark that the anisotropy of the in-plane effective masses m_M^* and m_K^* for wz-ZnO indicates that the corresponding bands are not completely parabolic throughout the **k** range used for the fitting. The symmetry properties of the effective-mass tensor would imply $m_M^* = m_K^*$ for strictly parabolic bands.

4.1 One-particle excitations

	wz-ZnO		rs-MgO	rs-CdO
$m_M^*(\Gamma_{7c})$	0.29	$m_X^*(\Gamma_{6c}^+)$	0.36	0.19
$m_K^*(\Gamma_{7c})$	0.39	$m_K^*(\Gamma_{6c}^+)$	0.42	0.25
$m_A^*(\Gamma_{7c})$	0.25	$m_L^*(\Gamma_{6c}^+)$	0.36	0.19
$m_M^*(\Gamma_{7+v})$	0.34	$m_X^*(\Gamma_{8v}^-)$	1.85	4.85
$m_K^*(\Gamma_{7+v})$	0.67	$m_K^*(\Gamma_{8v}^-)$	4.53	−1.35
$m_A^*(\Gamma_{7+v})$	2.47	$m_L^*(\Gamma_{8v}^-)$	3.21	−1.98
$m_M^*(\Gamma_{9v})$	2.45	$m_X^*(\Gamma_{8v}^-)$	1.61	2.33
$m_K^*(\Gamma_{9v})$	2.16	$m_K^*(\Gamma_{8v}^-)$	1.65	3.52
$m_A^*(\Gamma_{9v})$	2.45	$m_L^*(\Gamma_{8v}^-)$	2.37	−3.63
$m_M^*(\Gamma_{7-v})$	2.55	$m_X^*(\Gamma_{6v}^-)$	0.44	0.36
$m_K^*(\Gamma_{7-v})$	2.46	$m_K^*(\Gamma_{6v}^-)$	0.44	0.38
$m_A^*(\Gamma_{7-v})$	0.27	$m_L^*(\Gamma_{6v}^-)$	0.36	0.24

Table 4.3: Effective masses m^* (in units of the free-electron mass m) at the Brillouin zone center along the $\Gamma-M$, $\Gamma-K$, and $\Gamma-A$ directions for wz-ZnO and along the $\Gamma-X$, $\Gamma-K$, and $\Gamma-L$ directions for rs-MgO and rs-CdO. Values are given for the lowest conduction band and the three uppermost valence bands. For rs-MgO and rs-CdO the uppermost VB state is fourfold degenerate at Γ.

We find that in wz-ZnO the lowest CB and the Γ_9-derived VB are relatively isotropic with effective masses for the three high-symmetry directions not differing by more than 50%, whereas the two Γ_7-derived VBs show a pronounced anisotropy. The Γ_{7+} band is the light-hole band along $\Gamma-M$ and $\Gamma-K$ in the BZ and, therefore, in the plane perpendicular to the hexagonal c axis of the wz crystal. Its mass is roughly one order of magnitude larger in the direction parallel to the c axis. For the Γ_{7-}-derived band we find the opposite behavior.

For rs-MgO the anisotropy is much less pronounced and the two twofold degenerate VBs are heavy-hole related. Their masses are of the same order of magnitude in all three high-symmetry directions and are, roughly, one order of magnitude larger than the values for the light-hole band. The lowest CB turns out to be almost isotropic.

As with rs-MgO, the two uppermost VBs in rs-CdO are heavy-hole bands and the third VB is a light-hole one. The values in Table 4.3 show a remarkable difference to the case of MgO, since we find negative values for the effective mass of the uppermost two VBs along certain directions in **k** space. As discussed in Section 4.1.1 this can be traced back to the symmetry-forbiddance of the pd hybridization at the Γ point, explaining the convex curvature of the uppermost VBs along certain high-symmetry directions.

Experimentally determined effective masses

In experiment oftentimes quantities are measured that provide *inverse* effective masses. In addition, due to the respective experimental procedure, the components of the effective mass tensor are not necessarily accessed individually, instead average values are obtained. We take both aspects into account by deriving averages of inverse masses, i.e., *harmonic mean values*, for a comparison between our values (cf. Table 4.3) and experimental results.

For the lowest CB of wz-ZnO the harmonic mean value yields $m^* = 0.30\,m$ which compares well with values of $m^* = 0.28\,m$ [130] or $m^* = 0.29\,m$ [131] obtained from cyclotron-resonance experiments. M. Oshikiri et al. compared their measured values to theoretical results based on a conventional KS electronic structure, which yielded slightly smaller anisotropies and mass values [130] than we give in Table 4.3. From magneto-optical measurements [131] an effective mass of $m^* = 0.59\,m$ has been found for the Γ_{7+v} VB state perpendicular to the c axis. Though this value is slightly larger than the average $m^* = 0.45\,m$ of the inverse masses resulting from our calculations for the $\Gamma-M$ and the $\Gamma-K$ directions (cf. Table 4.3), our averaged value coincides with the result of a Zeeman-based measurement [131]. We calculated the effective mass of the CF split-off band Γ_{7-v} along $\Gamma-A$ as $m^* = 0.27\,m$. This is close to $m^* = 0.31\,m$, obtained from magneto-optical measurements [131], and agrees with another theoretical result calculated within DFT-LDA [125]. The harmonic mean value of the masses along $\Gamma-M$ and $\Gamma-K$ is $m^* = 2.50\,m$ and, hence, much larger than an experimental value, $m^* = 0.55\,m$ [131], for the mass perpendicular to the c axis but only twice as large as a value given by W. Lambrecht et al. [125]. Also, for the uppermost VB we find very good agreement for the effective masses and band anisotropies with Lambrecht's values derived from DFT-LDA [125]. In the case of the Γ_{9v} state we obtained large effective masses along all directions in **k** space, while the results in Ref. [125] confirm this only along $\Gamma-A$.

To our knowledge no measured values exist for the effective electron or hole masses in rs-MgO. Another theoretical study [132] reported a lowest isotropic CB, whereas we find a small anisotropy (cf. Table 4.3). In addition, our results do not indicate a degeneracy of the masses of the uppermost two VBs along the $\Gamma-X$ or $\Gamma-L$ directions and our values for the light-hole band are slightly larger. We attribute both to the neglect of QP corrections in Ref. [132].

For the effective mass of the CB of rs-CdO values in the range $m^* = 0.21\,m \ldots 0.3\,m$ have been obtained in experiments [133–135] and are in good agreement with our calculated values of $m^* = 0.19\,m$ or $m^* = 0.25\,m$ (cf. Table 4.3). No measurements for hole masses have been reported for this material. At least the fact that they are negative for certain directions in **k** space is consistent with the VBM at the L point for rs-CdO, which is experimentally confirmed [111].

4.1.3 Application: Band alignment at interfaces

In the proximity of an interface between two semiconductors we expect a certain "transition" region where the bulk band structures of the two materials merge into each other. Typically, band-edge discontinuities (i.e. VB and CB offsets) that are confined to only some atomic layers can occur, whereas band bending due to different doping profiles can extend further [136] (see Fig. 4.6). In the infinite bulk material only non-evanescent states, i.e.,

4.1 One-particle excitations

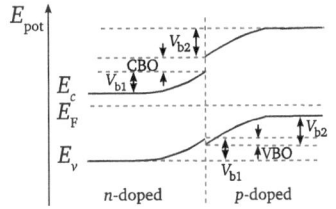

Figure 4.6: Schematic plot of the behavior of valence (E_v) and conduction (E_c) bands at the contact of an n-doped with a p-doped semiconductor. Band bendings (V_{b1} and V_{b2}) as well as valence- and conduction-band offsets (VBO, CBO) are indicated.

states with real **k** vectors, emerge. This constraint is no longer valid at interfaces where evanescent, exponentially decaying solutions can occur. These states are called *virtual gap states* (ViGS) or *interface-induced gap states* (IFIGS). Their eigenvalues occur within the gap and they are described by complex **k** vectors giving rise to a complex band structure [136]. We can envision these states as the continuation of bulk VB or CB states extending across the interface. At a certain energy, the branch-point energy (BPE) E_{BP}, they change their character from more VB-like (or acceptor-like) to more CB-like (donor-like). From the treatment of metal-semiconductor contacts the term *charge neutrality level* [137–139] is adopted sometimes.

In the absence of interface dipoles the band alignment can be achieved by aligning the BPEs of the two semiconductors that form the interface [136]. Interestingly, the BPE (also *effective mid-gap energy* in Ref. [138]) and, therefore, the band alignment, have entirely been traced back by J. Tersoff to the *bulk band structure* of the two semiconductors involved [138]. An obvious advantage of such an approximation is the complete neglect of the complicated and oftentimes unknown (structural) details of the interface. More specifically, we can access the E_{BP} of the three group-II oxides through our *ab-initio* results for the band structures of the bulk.

Method

Originally, Tersoff's method was based on the cell-averaged single-QP real-space Green's function [138]. Unfortunately, it turned out that it is extremely difficult in practical calculations to converge the Green's function with respect to the number of **k** points and CBs [140]. Besides this, Tersoff's method fails for systems where E_{BP} appears in the CB region, whereas there is experimental evidence for such a behavior: A branch-point energy within the first CB is assumed to be the origin of the electron accumulation e.g. at InN surfaces [141, 142].

Therefore, in this work we follow a different approximate approach to obtain E_{BP}. It is inspired by F. Flores *et al.* [137, 143] who estimated the BPE as the average of the mid-gap energies at the Γ, X, and L points of the BZ for face-centered cubic crystals. While another

	E_{BP} (theor.)	E_{BP} (exp.)	ΔE_c	ΔE_v
rs-MgO	5.42	–	2.07	−5.42
wz-ZnO	3.40	> 3.58 [146]	−0.19	−3.40
rs-CdO	2.45	1.30 ± 0.1 [147]	−1.77	−2.45

Table 4.4: Calculated and experimental values for the branch-point energies, E_{BP}, relative to the valence-band maximum. E_{BP} is used as the level of reference for the band offsets ΔE_c and ΔE_v. All values are given in eV.

early study [144] applied the concept of a BZ average, based only on the first of Baldereschi's special points to approximate the **k**-point sum, we rely on our band structures which are known with high accuracy for an entire set of **k** points, i.e. a Γ-centered MP [94] **k**-point mesh. Hence, we calculate the BPE by means of a complete BZ average using the expression:

$$E_{BP} = \frac{1}{2N_{KP}} \sum_{\mathbf{k}} \left[\frac{1}{N_{CB}} \sum_{i}^{N_{CB}} \varepsilon_{c_i \mathbf{k}}^{QP} + \frac{1}{N_{VB}} \sum_{j}^{N_{VB}} \varepsilon_{v_j \mathbf{k}}^{QP} \right] \quad (4.2)$$

For rs-MgO and rs-CdO only the lowest CB and the two uppermost VBs are included (see Ref. [145]). These numbers are doubled for wz-ZnO because the unit cell is twice as large. Though the choices of N_{CB} and N_{VB} significantly influence the results, the one for N_{CB} is clearly justified due to the huge dispersion of the lowest CB near Γ. We exclude the third VB (in cubic crystals) from the sum (4.2) due to its much larger **k** dispersion throughout the BZ compared to the two uppermost ones [145]. Overall, we estimate that the arbitrariness of choosing N_{CB} and N_{VB} introduces an uncertainty of up to 0.2 eV. We use the reliable bulk band structures of the group-II oxides calculated within this work (cf. Section 4.1.1) to compute E_{BP} from Eq. (4.2).

Branch-point energies and band discontinuities

In Table 4.4 we give our calculated results for the BPE jointly with measured values. In the case of wz-ZnO, the quantitative agreement with an experimental finding [146] as well as a calculated result, derived from a band alignment using hydrogen levels [139], is good. For rs-CdO $E_{BP} = (1.30 \pm 0.1)$ eV has been found by an experiment [147], which is somewhat lower than our calculated E_{BP} given in Table 4.4. In a different study of this material the uppermost VB state at Γ was used as the level of reference leading to $E_{BP} = (2.55 \pm 0.05)$ eV [148]. Taking the energy difference between the uppermost state at Γ and the VBM, 1.12 eV (cf. Section 4.1.1), into account we obtain $E_{BP} = 3.57$ eV, which again overestimates the experimental result.

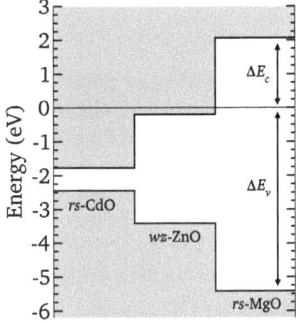

Figure 4.7: Conduction-band edges and valence-band edges for rs-MgO, wz-ZnO, and rs-CdO. The branch-point energies are used as the level of reference.

For wz-ZnO and rs-CdO we find that the CBM is below E_{BP} and attribute this fact to the low DOS close to the CBM. Despite the pronounced minimum of the first CB, which is 4 to 5 eV lower than the lowest CB states in the outer regions of the BZ for MgO, ZnO, and CdO (cf. Fig. 4.1), this band shows a strong dispersion. The weight of the outer regions of the BZ in the sum over **k**, Eq. (4.2), is much higher and, therefore, E_{BP} can occur within the lowest CB. A possible consequence of this situation is the high, unintentional n-type conductivity of nominally undoped ZnO or CdO surfaces. For rs-CdO there is evidence of an electron accumulation at the surface from X-ray photoemission spectroscopy and from angle-resolved photoemission spectroscopy [147, 149]. The agreement of the BPE and the CBM within 0.2 eV for wz-ZnO might be an indication why it can be used as transparent conductive oxide at least after doping with aluminum.

Using the BPE as a universal energy level of reference to align the energy bands of different semiconductors leads to the band lineups shown in Fig. 4.7. The offsets of the uppermost VB and the lowest CB, with respect to E_{BP}, can be interpreted as natural band discontinuities, ΔE_v and ΔE_c. From the results in this figure, we predict that a combination of the three oxides rs-MgO, wz-ZnO, and rs-CdO yields type-I heterostructures. In Ref. [145] we also successfully apply this approach to In_2O_3 and three nitrides.

4.2 Two-particle excitations

4.2.1 Impact of many-body effects on the optical properties

In this section we illustrate the influence of QP effects, as well as excitonic and local-field effects, on the optical properties of MgO, ZnO, and CdO. We start by comparing three different levels of many-body perturbation theory by employing (i) the independent-particle approx-

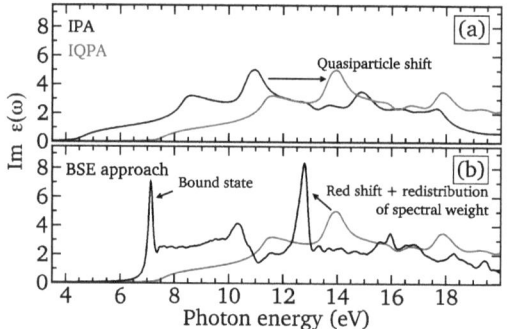

Figure 4.8: Imaginary parts of the dielectric function of rs-MgO calculated within the independent-particle approximation (black) are compared to results within the independent-QP approximation (gray) in (a). The black curve in (b) shows the result which has been calculated using the BSE approach.

imation (IPA), i.e., eigenvalues and wave functions from DFT-GGA, or (ii) the IQPA which involves wave functions from DFT-GGA+U together with QP energies (simulated by Δ, cf. Section 3.1.2) to calculate the DF of non-interacting electron-hole pairs. Finally, (iii) excitonic and local-field effects are incorporated by solving the BSE for the optical polarization function (based on the IQPA as the starting electronic structure).

In Fig. 4.8(a) the influence of QP corrections becomes clear from the comparison of the imaginary part of the DF of rs-MgO obtained using the IPA to the one calculated within IQPA. The inclusion of QP energies leads to a blueshift on the order of 1...3 eV for the oxides studied in this work (cf. Section 4.1). The influence of excitonic and local-field effects becomes clear from a comparison of the IQPA curve with the result calculated using the BSE approach in Fig. 4.8(b). The electron-hole interaction causes a redshift of the BSE curve towards lower photon energies, with respect to the IQPA result. This redshift does not compensate the QP blueshift, thus an overall blueshift of the BSE spectrum of about 1...2 eV with respect to the IPA curve remains. In addition, a strong enhancement of the peak intensities and plateau heights due to the Coulomb interaction is visible at low energies in Fig. 4.8(b). Both the redshift and the Coulomb enhancement are referred to as redistribution of oscillator strength caused by the excitonic effects. In spite of this effect, peaks of the BSE curve can be related to structures in the IQPA spectrum – except for the remarkable feature at the absorption onset; it is attributed to a bound, Wannier-Mott-like electron-hole-pair state with large oscillator strength, an *exciton*. As the lowest optical excitation of the system, this peak originates from the lowest eigenstates of the excitonic Hamiltonian with excitation energies smaller than the QP gap.

4.2.2 Complex frequency-dependent dielectric function

The discussion in the previous section elucidated the impact of the many-body effects on the DF of the group-II oxides and exemplified how theoretical calculations go beyond merely reproducing or predicting experimentally accessible quantities. Moreover, they provide insight into the underlying physics by disentangling different effects which contribute to the final result.

In the case of the optical properties, such as the DF, an interpretation of the respective peaks in terms of the involved VBs and CBs is helpful. Frequently, the mapping is done by assigning the spectral features merely to van Hove singularities at high-symmetry points in the band structure. This procedure is already arguable when interpreting the DF calculated within the IPA, at least for the oxides studied in this work, because we found that the corresponding transitions oftentimes must be associated with larger regions of the BZ. Besides this, taking the Coulomb interaction between electrons and holes into account additionally couples VB and CB states from different **k** points. This Coulomb-induced mixing can render it even less feasible to identify certain high-symmetry **k** points as an origin for certain spectral features.

Therefore, in this work we combine two approaches to analyze remarkable spectral features of the DFs. Comparing the IQPA result to the BSE curves enables us to assign structures in the DF from the BSE approach to peaks in the independent-QP spectrum. In a second step, we distinguish between contributions to the IQPA spectrum that are caused by a high joint DOS or by matrix-element effects. This allows us to trace back certain peaks to band complexes that particularly contribute to the respective transitions.

Results and discussion: MgO

In Figs. 4.9(a) and (b) we plot the result for the real and imaginary part of the DF calculated using the BSE approach. Comparison to a curve measured using spectroscopic ellipsometry [150] proves excellent agreement regarding the energetic positions of the peaks in the imaginary part. Also, two DFs derived from reflectance measurements [151, 152] by means of Kramers-Kronig analysis reveal only slight deviations of peak positions and intensities. For photon energies above 12 eV, Fig. 4.9 displays several intensity deviations, which may be partly related to a larger lifetime and instrumental broadening for transitions in that energy range, i.e., with final states above the vacuum level. In addition, we slightly underestimate the peak positions above 15 eV, which we attribute to the missing energy dependence of the QP corrections when merely a scissors operator is used (cf. discussion in Section 4.1.1). Comparing to older calculations that also include the electron-hole interaction [59, 153], we find that our more converged results agree better with measured curves.

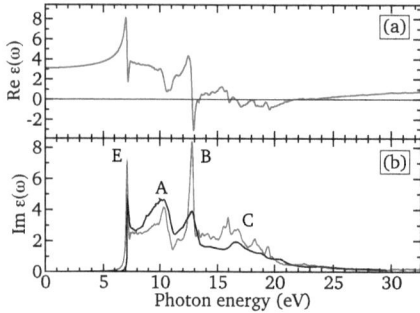
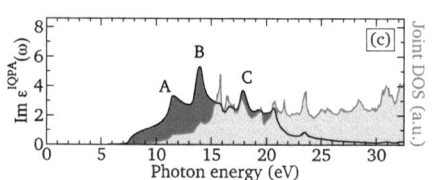

Figure 4.9: Real (a) and imaginary (b) parts of the dielectric function of rs-MgO, including excitonic and local-field effects (gray curves), together with an experimental result (black curve) from Ref. [150]. For additional comparison the imaginary part of the dielectric function calculated within independent-particle approximation (black curve) and the joint density of states (gray curve) are plotted (c).

For a deeper analysis we study the DF calculated within IQPA, as well as the joint DOS, in Fig. 4.9(c). First of all, this points out the strong modification due to the electron-hole interaction, especially in the band-edge region, as well as the strong Coulomb-induced spectral redistribution that we discussed before (see Section 4.2.1). A comparison of the two curves in Fig. 4.9(c) reveals the strong influence of the optical transition-matrix elements. While the energetic positions, e.g. of peaks A and B, can be matched to structures in the joint DOS, we find that the line shape of the independent-QP spectrum hardly resembles that of the joint DOS, proving the large impact of the optical transition-matrix elements. In the same figure, the assignment of the peaks of the independent-QP spectrum and the DF which includes excitonic effects is pointed out by the labels. Investigating the contributions to the IQPA spectrum indicates that peak A mainly can be attributed to transitions between the two highest VBs and the CBs, whereas peak B is almost entirely composed of transitions from the uppermost VB into the CBs. Peak C originates mainly from transitions of the second and the third VB.

Results and discussion: ZnO

Our result for the DF of wz-ZnO, calculated using the BSE approach, is compared in Fig. 4.10 to an experimental spectrum obtained by means of spectroscopic ellipsometry [154]. We find good agreement for the peak positions, not only of the bound excitonic state at the absorption edge (E), but also at higher energies. Figure 4.10 shows the slight underestimation of the energetic position of peak A at around 8.9 eV and above, which we attribute to merely having used a scissors operator Δ for the QP corrections [see also the differences in Fig. 4.1(b)], as

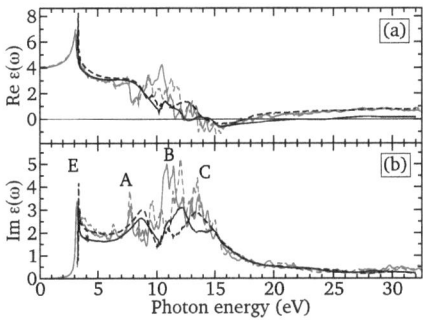
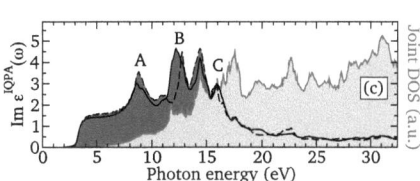

Figure 4.10: Real (a) and imaginary (b) parts of the dielectric function of wz-ZnO, including excitonic and local-field effects (gray curves), together with an experimental result (black curves) from Ref. [154]. For comparison, the imaginary part of the dielectric function calculated within independent-particle approximation (black curve) and the joint density of states (gray curve) are plotted (c). While solid curves correspond to ordinary light polarization, the dashed curves represent extraordinary polarization.

already discussed for rs-MgO. We marginally overestimate the plateau height in the energy region around $\hbar\omega \approx 4\ldots 7$ eV, which might be an artifact of the surface-layer corrections used in the description of the ellipsometry measurements. The calculated and the measured curve agree in finding the optical anisotropy due to the hexagonal crystal structure to be very small at photon energies between peaks E and A, as well as above 15 eV [cf. Fig. 4.10(a) and (b)]. Conversely, it is more pronounced between 8 eV and 15 eV. This is also confirmed by another measured curve which was derived via Kramers-Kronig analysis of reflectivity data [155]. In addition, our result is better converged than an earlier calculation of the DF including excitonic effects [156], where not even enough CBs for a calculation of the DF up to 15 eV were included.

Investigating the independent-QP spectrum in Fig. 4.10(c) confirms the expected large impact of excitonic effects on the DF of wz-ZnO. More importantly, this figure shows that the influence of the optical transition-matrix elements is stronger for wz-ZnO than in the case of rs-MgO. The shape of peak A and the region around peak C is remarkably modified. Our analysis shows that the peak structure A consists mostly of transitions from the uppermost four VBs into the CBs. The broad peak complex B, at higher energies between 10 to 15 eV, mainly originates from transitions from all O $2p$ VBs into the CBs. Interestingly, above photon energies of about 20 eV, roughly $20\ldots 50\,\%$ of the imaginary part of the DF arise from transitions originating in the Zn $3d$ states. Besides this, we are able to trace the optical anisotropy back to the uppermost three O $2p$ VBs which contribute most to this effect. While transitions

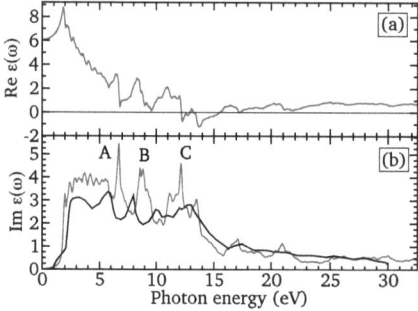
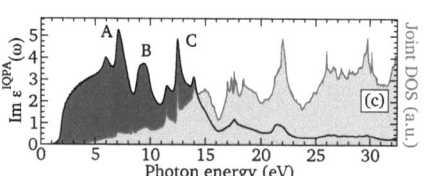

Figure 4.11: Real (a) and imaginary (b) parts of the dielectric function of rs-CdO, including excitonic and local-field effects (gray curves), together with an experimental result (black curve) from Ref. [157]. For additional comparison the imaginary part of the dielectric function calculated within independent-particle approximation (black curve) and the joint density of states (gray curve) are plotted (c).

from the third VB cause the large contributions between roughly 10...11.5 eV for perpendicular light polarization, transitions mainly from the first and second VB form the peaks for parallel polarization between 12...14 eV. As for the imaginary part, also the real part of the DF agrees well with the measured curve with the largest deviations between photon energies of 9...14 eV.

Results and discussion: CdO

In Figs. 4.11(a) and (b) we show the curves for the real and the imaginary parts of the DF, calculated using the BSE approach. We compare the imaginary part to an experimental result obtained by means of Kramers-Kronig analysis of reflectance data [157]. While the agreement is good up to photon energies of about 6 eV, we find again the aforementioned underestimation of the peaks' energetic positions at higher energies due to the lacking energy dependence of the scissors operator. An estimate of this effect from the band structure of rs-CdO in Fig. 4.1(c) explains deviations on the order of about 1...2 eV. Evidently, the indirect semiconductor rs-CdO does not show the pronounced peak close to the absorption edge that we attributed to a bound excitonic state in the case of rs-MgO and wz-ZnO. Aside from the indirect gap of rs-CdO, the much stronger screening (see next page) in this material has also been spotted as a reason for this behavior. In a two-band Wannier-Mott (WM) model [158] [cf. Eq. (5.3)], the exciton-binding energy is inversely proportional to the square of the static dielectric constant. This dependence points out that the excitonic effects strongly decrease with an increase of the screening of the electron-hole interaction and, consequently,

4.2 Two-particle excitations

	rs-MgO	wz-ZnO		rs-CdO
IPA [19]	3.16	$\varepsilon_\infty^\parallel = 5.26$	$\varepsilon_\infty^\perp = 5.24$	7.20
IQPA	2.77	$\varepsilon_\infty^\parallel = 3.64$	$\varepsilon_\infty^\perp = 3.58$	5.52
BSE	3.12	$\varepsilon_\infty^\parallel = 4.08$	$\varepsilon_\infty^\perp = 4.01$	6.07
Exp. [109]	2.94	$\varepsilon_\infty^\parallel = 3.75$	$\varepsilon_\infty^\perp = 3.70$	3.80 ... 7.02

Table 4.5: Electronic static dielectric constants of the three group-II oxides, calculated within the independent-particle approximation (IPA), independent-QP approximation (IQPA), and including excitonic and local-field effects (BSE). Experimental values are given for comparison.

we expect the impact of the excitonic effects to be relatively small for rs-CdO.

This expectation is further confirmed by a comparison of the BSE result to the IQPA curve in Fig. 4.11(c) which shows that the two curves look more alike than those of rs-MgO or wz-ZnO. Besides this, for rs-CdO the influence of the optical transition-matrix elements is smaller and the IQPA spectrum resembles the joint DOS, especially in the energy range below 13 eV. Analyzing the contributions to the DF within the IQPA reveals that peak A consists mostly of transitions from the uppermost VB into the CBs, whereas peak B is composed of equal contributions from all three uppermost O $2p$ VBs. The peak complex C can be clearly related to a high joint DOS [cf. Fig. 4.11(c)] and the two peaks around $\hbar\omega = 17$ eV and $\hbar\omega = 21$ eV are attributed to transitions from the Cd $4d$ states.

Static dielectric constants

By computing converged results for the DF, using the BSE approach for low as well as high photon energies, and merging the two (cf. Section 3.2) we access the real part of the DF and, therefore, the static *electronic* dielectric constant $\varepsilon_\infty = \mathrm{Re}\,\varepsilon(\omega = 0)$. To exclude contributions from phonon excitations, it is derived as the high-frequency limit (with respect to phonon frequencies) from measurements, which is a possible source for experimental uncertainties. In Table 4.5 we compare the values for ε_∞, calculated within the IPA, the IQPA, and from the solution of the BSE, to measured values and obtain good agreement of the BSE results with deviations lower than 10% for all three oxides. For wz-ZnO we can even confirm that the parallel component of ε_∞ is slightly larger than the value for perpendicular light polarization.

Comparing the results arising from the IPA and the IQPA demonstrates the influence of the QP corrections on ε_∞, whereas the corrected d-band positions that enter the IQPA have almost no impact on the DF, as discussed in the previous section. An opening of the gap leads to a shrinkage of the static dielectric constant due to the Kramers-Kronig relation, which states that a smaller band gap is necessarily related to a larger static dielectric constant. This explains our findings for all three oxides and also the chemical trend of an increasing value of ε_∞ along the row rs-MgO, wz-ZnO, rs-CdO (cf. Table 4.5).

In fact, the difference between the dielectric constants calculated within IQPA and the experimental values indicates that a correct description of the band gap is not enough. We find that the IQPA results underestimate the experimental ones, whereas the BSE values agree much better. Since the influence of the electron-hole interaction on the gap is relatively small, we attribute this improvement to the redistribution of oscillator strength that we pointed out before.

So far, only the *electronic* contributions to the static dielectric constant have been discussed. In the low-energy region phonon effects are also important and the DF of the material can strongly deviate from the DF emerging merely from the electronic contributions. Consequently, the values for ε_s, as the static dielectric constants, given in the literature for the three oxides differ noticeably from ε_∞. While $\varepsilon_s = 9.8$ is found for rs-MgO and $\varepsilon_s^\parallel = 8.75 / \varepsilon_s^\perp = 7.8$ for wz-ZnO, the deviations are largest in the case of rs-CdO where a value as large as $\varepsilon_s = 21.9$ is reported [131]. Of course, this significant contribution to the screening has a large impact on the electron-hole interaction. In materials where the *lattice dynamics* of the screening become important, the use of only the electronic static dielectric constant is questionable [159]. In Ref. [125] a Pollmann-Büttner model has been used to tackle this problem, which was more successful than merely using experimental results for ε_∞ to screen the electron-hole interaction. Since it is not yet entirely clear how the electron-lattice interaction can be consistently included in the *ab-initio* approach used in this work, we have restricted ourselves to the use of the IPA values of ε_∞ calculated within the GGA+U approximation (cf. Appendix A.2) to screen the electron-hole interaction in the BSE approach. Since these ε_{eff} are between ε_s and ε_∞ we benefit from a certain cancellation of errors that consequently occurs.

4.2.3 Excitons and spin-orbit coupling

The lowest eigenstates of the excitonic Hamiltonian describe excitons with a binding energy E_B which is defined as the difference of the energy of the non-interacting electron-hole pair and the respective eigenvalue. For the oxides studied in this work, values of E_B on the order of about 60 meV for wz-ZnO [131] or 80 meV (145 meV) for rs-MgO [152] (Ref. [160]) have been derived from measurements. Only due to the adaptive **k**-point sampling scheme (cf. Section 3.2) are we able to achieve the calculation [61, 90] of converged values for E_B. However, the appropriate description of the screening is difficult for reasons elucidated in the

4.2 Two-particle excitations

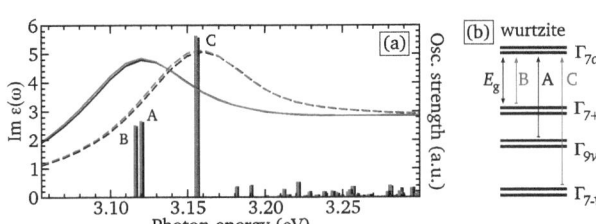

Figure 4.12: Imaginary part of the dielectric function (curves) of wz-ZnO in the vicinity of the absorption edge together with the lowest eigenvalues of the excitonic Hamiltonian and the respective oscillator strengths (bars). In subfigure (b) the nomenclature of the excitons A, B, and C is explained.

preceding section.

Here, we focus on the effect of the spin-orbit interaction on the lowest optical excitations. Using the irreducible representations of the uppermost three VB states and the lowest CB state at the Γ point, as given in Fig. 4.4, we derive the allowed optical transitions, along with their polarization dependence, by means of group theory. In the respective multiplication table [161] for the C_{6v}^4 symmetry group of the wz structure we find that the lowest CB state (Γ_7 symmetry) and a VB state (with Γ_7 symmetry) lead to $\Gamma_7 \times \Gamma_7 \to \Gamma_5 + \Gamma_1 + \Gamma_2$, whereas we obtain $\Gamma_7 \times \Gamma_9 \to \Gamma_5 + \Gamma_6$ for Γ_9-type VBs. By means of the irreducible representation of the dipole operator for this group, it turns out that of the terms in these sums, only Γ_5 (Γ_1) is dipole-allowed for perpendicular (parallel) light polarization. Our calculations indicate that the Γ_5-related transitions mainly originate from the Γ_{9v} (Γ_{7+v}) VBs and we denote them as A (B) excitons. The C exciton is associated with Γ_1-derived transitions, mainly from the Γ_{7-v} VB [cf. Fig. 4.12(b)].

Performing the same analysis for the rs crystal structure with the multiplication table of the O_h^5 group one finds two (cf. Fig. 4.4) different products, $\Gamma_{6+} \times \Gamma_{8-} \to \Gamma_{12'} + \Gamma_{15} + \Gamma_{25}$ and $\Gamma_{6+} \times \Gamma_{6-} \to \Gamma_{2'} + \Gamma_{15}$, out of which only the Γ_{15}-related transitions are dipole-allowed.

In the following, we revisit the problem of the VB-ordering in wz-ZnO which we tackled in Section 4.1.2 by studying the respective QP energies. Taking the electron-hole interaction into account leads to the picture in Fig. 4.12(a) where we show the imaginary part of the DF in the direct vicinity of the absorption edge, along with the eigenvalues and oscillator strengths of the eigenstates of the excitonic Hamiltonian that are found in this energy region. The spin-orbit interaction is included by means of the approach described in Section 3.2.2. The small but non-vanishing splittings of the A-, B-, and C-exciton-related peaks [cf. Fig. 4.12(a)] arise due to a larger influence of the spin-orbit interaction apart from the Γ point. In addition, the spin-orbit induced and the CF-related splittings into the A, B, and C excitons are clearly visible in Fig. 4.12, along with the polarization anisotropy which is in accordance with our discussion based on group theory: The lowest four eigenvalues (A and B excitons) are visible in perpendicular polarization and the next two lowest eigenvalues (C exciton) occur for parallel polarization. In Fig. 4.12 the small energy differences of the absorption onsets that arise from this polarization anisotropy are clearly visible. However, the electron-hole interaction does not change the ordering of the lowest optical transitions with respect to the ordering of the states in the QP band structure. Due to the differences of the exciton-binding energies (cf. Ref. [90]), the splitting between A and B (A and C) amounts to -12.1 meV (44.2 meV), which is 0.8 meV more (4.1 meV less) than the difference of the respective QP energies (cf. Table 4.2). Estimates based on Eq. (5.3) indicate that when the screening is as large as $\varepsilon_s \approx 9$ the effect of the electron-hole interaction is reduced to less than 1 meV.

4.2.4 Application: Electron-energy loss function

All linear-optical properties can be derived from the complex DF as a response function of the system. Though we calculated the Fresnel reflectivity $R(\omega)$ in Ref. [89], we want to focus in this work on the energy loss of an electron that is scattered by a sample. When treating the electron as a classic particle, in the non-relativistic limit the energy loss of the electron can (in the limit of vanishing transferred momenta) be described by the electron-energy loss function,

$$-\operatorname{Im} \varepsilon^{-1}(\omega) = \frac{\operatorname{Im} \varepsilon(\omega)}{(\operatorname{Re} \varepsilon(\omega))^2 + (\operatorname{Im} \varepsilon(\omega))^2}, \quad (4.3)$$

where $\hbar\omega$ denotes the loss energy. Equation (4.3) neglects retardation and surface effects.

We plot our calculated results for rs-MgO, together with a measured curve by S. Kohiki et al. [162], in Fig. 4.13(a). Although the experimental curve does not show any fine structure, we find a good agreement for the overall shape as well as for the position of a pronounced plasmon resonance centered around ≈ 23 eV. By means of the relation

$$\hbar\omega_p = \hbar\sqrt{\frac{e^2}{\varepsilon_0 m} \cdot n} \quad (4.4)$$

we can relate the cell-averaged electron density $n = N/\Omega_0$ to the plasma frequency ω_p. Taking the O $2s$ and O $2p$ electrons into account we obtain a value of $\hbar\omega_p = 23.9$ eV which agrees well with a structure in Fig. 4.13(a). It has to be pointed out that, probably due to sample-quality-related effects such as impurities or defects, the experimental onset of the loss function appears at roughly 5 eV, which is below our onset at about 7.2 eV.

In Fig. 4.13(b) we compare our result for wz-ZnO to a measurement by J. L. Freeouf et

4.2 Two-particle excitations

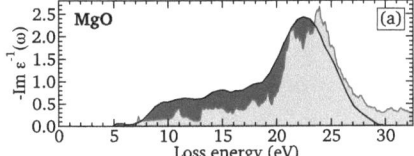
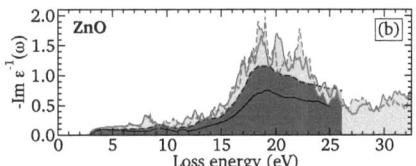
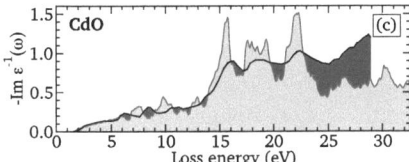

Figure 4.13: Electron-energy loss function $-\text{Im}\,\varepsilon^{-1}(\omega)$ of rs-MgO (a), wz-ZnO (b), and rs-CdO (c) including excitonic effects (gray curves). We compare to experimental results (black curves) from Ref. [162] (MgO) and Ref. [157] (ZnO, CdO). For the hexagonal wz-ZnO curves for the ordinary direction (solid) as well as for the extraordinary direction (dashed) are shown.

al. [157] and obtain a good overall agreement of the curve shape. Between 15 and 25 eV a broad plasma resonance occurs which is related to small values of the real part of the DF between 18 and 23 eV, with a zero at around 17 eV. Using Eq. (4.4) we find plasma frequencies of 10.5 eV, 18.3 eV, or 23.6 eV when *only* contributions from the O 2s, O 2p, or Zn 3d electrons are taken into account. Comparing these values to the plot in Fig. 4.13(b) indicates that three distinct contributions can barely be observed and, instead, a mixing of the contributions occurs, leading to the broad plasma resonance mentioned before. In addition, the underestimation of the energetic positions of peaks that has been discussed for the DF (cf. Section 4.2.2) also carries over to the description of the loss function. We distinguish between $\parallel$ and $\perp$ quantities, following the previously introduced definition, and find a relatively small anisotropy for the entire curve. This is not confirmed by the measurement [157] shown in Fig. 4.13(b), but more so by a result obtained via Kramers-Kronig analysis of reflectivity data [155].

The calculated result for the electron-energy loss function of rs-CdO is shown in Fig. 4.13(c) along with the measured curve of J. L. Freeouf *et al.* [157]. In this case we do not obtain one single plasma resonance structure for the *s*, *p*, or *d* electrons, but a clear three-peak structure at energies between 15 and 23 eV. This behavior is confirmed by the experimental result despite the underestimation of the energetic positions of the peaks of the calculated curve that has already been discussed. By means of Eq. (4.3) we compute plasma frequencies of 10.1 eV, 17.4 eV, or 22.5 eV caused by the O 2s, O 2p, or Cd 4d electrons and can relate them to structures in Fig. 4.13(c). The peak at about 18 eV cannot be assigned to *s*, *p*, or *d* electrons with this type of analysis and arises, most likely, due to a combination of these states.

4.3 Summary

In this chapter we provided a detailed analysis of the electronic structure and the optical properties of the group-II oxides rs-MgO, wz-ZnO, and rs-CdO using an *ab-initio* description.

We compared the QP energies and the DOS that we obtained by means of the sophisticated HSE+G_0W_0 approach to experimental results and found reassuring agreement for all three oxides. Furthermore, our results have been used to derive natural band discontinuities, leading to the conclusion that a combination of rs-MgO, wz-ZnO, and rs-CdO yields type-I heterostructures. In addition, we proved that a mapping onto the computationally less expensive GGA+U+Δ method yields starting electronic structures that are suitable for calculating the excitonic Hamiltonian. The influence of the spin-orbit coupling has been taken into account and the ordering of the uppermost VB states was investigated.

In Section 4.2 the complex frequency-dependent DFs of rs-MgO, wz-ZnO, and rs-CdO were presented and interpreted. We found a remarkable influence of the electron-hole interaction by comparing different levels of the many-body perturbation theory. Besides this, the impact of spin-orbit coupling on the lowest eigenstates of the Hamiltonian was investigated and the electron-energy loss function was derived from the DF. We compared our results to measured curves as far as they are available. We also found that the experimentally observed splittings of the uppermost VB states agree well with our findings when the electron-hole interaction is included in the theoretical description.

5 Lattice distortions: Strain and non-equilibrium polymorphs

> Without deviation from the norm, progress is not possible.
>
> Frank Zappa

In order to gain a thorough understanding of the properties of a material by experimentation it is undoubtedly helpful to study pure, ideal crystals or samples with as few defects as possible. Accordingly, physical as well as chemical techniques for their preparation have been improved continuously and, nowadays, single crystals of very good quality are available for *rs*-MgO as well as for *wz*-ZnO. In the preceding chapter the electronic and optical properties have been studied extensively for their equilibrium polymorphs. However, occasionally systems of reduced dimensionality attract even more interest than the simple bulk materials since they come along with interesting and, with respect to bulk materials, *new* physical effects. In the context of nanoscience, therefore, thin films, small crystallites, and an entire variety of nanostructures, which demand their own distinct preparation techniques, are investigated.

Along these lines, thin films are somewhat outstanding since they are particularly important when the fabrication of large crystals of a material is difficult to achieve. They can be fabricated via deposition on various substrates using different methods. Depending on the magnitude of the *lattice mismatch* between the substrate and the film, such a procedure can lead to the presence of unintended strains in the sample. Systematic experimental studies of the behavior of the electronic structure in the presence of strain do exist, for instance, for *wz*-ZnO [123, 124]. Thus, in this chapter we investigate the influence of uniaxial as well as biaxial strain on the uppermost VB states, the DF, and the exciton binding energies of this material.

Moreover, when the crystal structures of the substrate and the deposited material differ, the film might, due to the growth, even adopt the lattice structure of the underlying substrate within several atomic layers. For such a strong deviation from the equilibrium atomic geometry one cannot expect that the influence on the electronic and optical properties is negligible. Since wz-ZnO is readily available as a substrate, we examine the electronic band structure and the optical properties of MgO and CdO, assuming that they occur in a non-equilibrium wz structure when deposited as thin films on wz-ZnO. Both inherent strain and non-equilibrium structures exist and occur to some extent also in alloys and heterostructures.

5.1 Uniaxial and biaxial strain in ZnO

In Section 4.1.2 of the preceding chapter the ordering of the uppermost three VB states (cf. Fig. 4.4) in wz-ZnO was discussed and compared to experimental results. While the deviations that we found between differences of our calculated QP energies and measured VB splittings were traced back to the influence of excitonic effects in Section 4.2.3, we want to investigate in the following to which extent possible uniaxial (parallel to the crystal's c axis) or biaxial (perpendicular to the crystal's c axis) strains in a sample might be responsible for changing the (relative) energetic positions of the energy levels at the VBM that are split by the CF or the spin-orbit coupling.

For the uniaxially and biaxially strained cells, the ground-state total energies are determined from the minimum of $E(V)$ curves that are calculated within the DFT-GGA. Subsequently, the relaxed atomic coordinates are obtained by minimizing the forces on the ions. To incorporate uniaxial strain into these calculations we fix the c lattice constant (cf. Fig. 4.1), whereas a is allowed to relax. Contrary, the a lattice constant is fixed and c is relaxed when accounting for biaxial strain. Using the equilibrium values a_0 and c_0 the uniaxial strain is defined as $\varepsilon_u = (c - c_0)/c_0$, whereas for biaxial strain it holds that $\varepsilon_b = (a - a_0)/a_0$. We are studying two compressive strains ($\varepsilon_x = -0.02$ and -0.01) and two tensile strains ($\varepsilon_x = 0.01$ and 0.02) in both cases (x = {u,b}). By means of these *ab-initio* calculations we gain insight into the properties of wz-ZnO even beyond the experimentally accessible conditions since, for this material, the typical strain values that can be achieved without destroying the samples are by roughly a factor of 10 smaller than what we study in this work [163].

We calculate the electronic structures for the different strained lattice geometries by means of the HSE+G_0W_0 approach, including SOC (cf. Sections 2.3 and 3.1). Using the expression

$$A_x(Z) = \partial Z(\varepsilon_x)/\partial \varepsilon_x|_{\varepsilon_x=0}, \qquad x = \{u,b\}, \tag{5.1}$$

we derive the strain coefficients A_x for quantities Z, such as gaps and VB splittings. In addition, for the biaxially strained cells we also compute the optical properties by solving the BSE (cf. Sections 2.4 and 3.2).

5.1 Uniaxial and biaxial strain in ZnO

	quantity	A_u	A_b
without SOC	$E_g = \varepsilon^{QP}(\Gamma_{1c}) - \varepsilon^{QP}(\Gamma_{5v})$	−3.91	−0.37
	$\Delta_1^{\text{no SOC}} = \varepsilon^{QP}(\Gamma_{5v}) - \varepsilon^{QP}(\Gamma_{1v})$	3.19	−5.08
with SOC	$\varepsilon^{QP}(\Gamma_{9v}) - \varepsilon^{QP}(\Gamma_{7+v})$	0.17	−0.07
	$\varepsilon^{QP}(\Gamma_{9v}) - \varepsilon^{QP}(\Gamma_{7-v})$	3.26	−5.09
	Δ_1^{qc}	3.21	−4.62
	$\Delta_2^{qc} = \Delta_3^{qc}$	0.05	−0.05

Table 5.1: Linear uniaxial A_u and biaxial A_b strain coefficients (in eV) for the gap, the VB energy splittings, and the derived quantities Δ_1, Δ_2, Δ_3 of wz-ZnO.

5.1.1 Quasiparticle energies in the proximity of the band gap

The ordering of the uppermost VB states in unstrained wz-ZnO has been found to be Γ_{7+v}–Γ_{9v}–Γ_{7-v} in Section 4.1.2, with splittings between these levels as given in Table 4.2. In the presence of uniaxial or biaxial strain, we obtain the picture shown in Fig. 5.1 for the QP energies around the fundamental gap at Γ and give the strain coefficients for the splittings of the VBs, as calculated from Eq. (5.1), in Table 5.1. The plot of the QP energies indicates a remarkable impact of the uniaxial strain on the lowest CB, whereas the influence of biaxial strain is comparably small. Accordingly, the uniaxial deformation potential (cf. Table 5.1), −3.91 eV, is about ten times larger than the value of −0.37 eV for the biaxial deformation potential.

For the VB states we studied the strain dependence of the energy splittings and of the **k·p** parameters (within the quasi-cubic approximation, i.e., Δ_1^{qc} and $\Delta_2^{qc} = \Delta_3^{qc}$). The cases when Eq. (4.1) yields imaginary values for $\Delta_2^{qc}/\Delta_3^{qc}$ (cf. Section 4.1.2), i.e., $\varepsilon_u = -0.02$ and $\varepsilon_b = 0.01$, were excluded from the linear fits to determine the deformation potentials according to Eq. (5.1). For both, uniaxial as well as biaxial strain, there is an influence on the spin-orbit splitting as can be seen from the clearly non-vanishing deformation potentials for $\Delta_2^{qc}/\Delta_3^{qc}$ or those for the energy difference between the Γ_{7+v} and the Γ_{9v} state. Comparing the linear strain coefficients for $\Delta_2^{qc}/\Delta_3^{qc}$ to experimental values [123] (measured for hydrostatic pressure) shows the same order of magnitude, though they deviate from the values given in Ref. [124]. Contrary, for the CF split-off level Γ_{7-v} we observe from Figs. 5.1(a) and (b) a much larger and almost linear decrease of its energetic position with the applied uniaxial strain, and an even larger increase for biaxial strain. Correspondingly, we find the deformation potential for Δ_1^{qc}, as well as that for the energy difference between the Γ_{7-v} and the Γ_{9v} state, to be roughly one order of magnitude larger than in the case of the uppermost two valence states (cf. Table 5.1). By means of the expression $A' = A/Y$ and using the biaxial modulus $Y = 216$ GPa [90] we can relate the strain coefficient of the CF split-off level Γ_{7-v} (cf. Table 5.1) to the biaxial stress coefficient. We obtain $A' = -2.35$ meV/kbar which agrees well with a measured value of −1.93 meV/kbar [124].

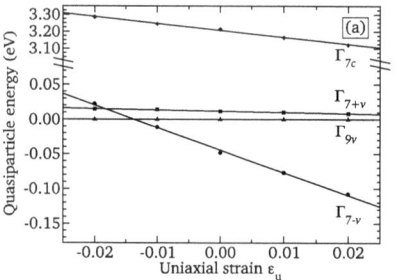 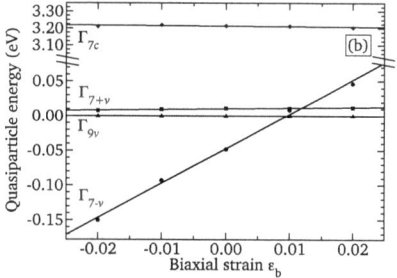

Figure 5.1: Quasiparticle energies at the Γ point obtained from the HSE+G_0W_0 approach, including the spin-orbit interaction, plotted versus uniaxial (a) or biaxial (b) strain. The Γ_{9v} level is taken as energy zero.

The QP band structure of wz-ZnO, shown in Fig. 5.1, demonstrates that both a compressive uniaxial or tensile biaxial strain as large as approximately 2 % lead to a change in the band ordering since the CF split-off level then becomes the uppermost VB. On the other hand, even for these large strains the ordering of the uppermost two valence states (Γ_{7+v} and Γ_{9v}) does not change.

5.1.2 Excitons under the influence of biaxial strain

In Section 4.2.3 it has been pointed out that the electron-hole interaction exerts an influence on the splittings of the uppermost VB states when these splittings are derived from optical properties. In the following, we want to extend these investigations by taking biaxial strain into account. This is of practical relevance for samples that were grown along the direction of their c axis on a substrate which is not completely lattice-matched. We adopt the nomenclature for the excitons as introduced previously [cf. Fig. 4.12(b)], i.e., A ($\Gamma_{9v} \rightarrow \Gamma_{7c}$), B ($\Gamma_{7+v} \rightarrow \Gamma_{7c}$), and C ($\Gamma_{7-v} \rightarrow \Gamma_{7c}$).

In this section we focus on the energetic distance between the CF split-off level (Γ_{7-v}) and the two uppermost VB states (Γ_{7+v} and Γ_{9v}), since in the preceding section (cf. Fig. 5.1) we observed that the splitting between the Γ_{7+v} and the Γ_{9v} level depends only weakly on strain. Therefore, we do not need to resolve the small splitting between these two states when plotting the imaginary part of the DF in Fig. 5.2. This significantly reduces the computational effort since we solve a BSE for the respective atomic structure of each strained unit cell separately so as to investigate the strain dependence of the DF. The respective static dielectric constant which determines the screening of the electron-hole interaction is adopted for each cell as well. We find a linear increase of its value (computed using the GGA+U approach), go-

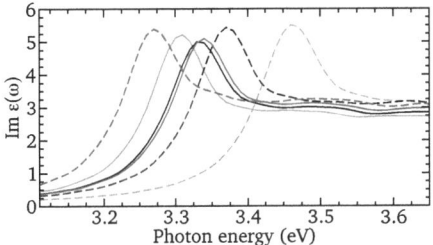

Figure 5.2: Imaginary part of the dielectric function versus photon energy (in the vicinity of the absorption edge) for a biaxial strain of $\varepsilon_b = -0.02$ (light gray curves), $\varepsilon_b = 0.02$ (dark gray curves), and the unstrained case (black curves). We distinguish between ordinary (solid lines) and extraordinary (dashed lines) polarization.

ing from $\varepsilon_{\text{eff}} = 4.29$ ($\varepsilon_b = -0.02$) to $\varepsilon_{\text{eff}} = 4.49$ ($\varepsilon_b = 0.02$) after averaging over all polarization directions.

From the different BSEs we calculate the DFs for different amounts of biaxial strain in wz-ZnO. In the resulting plot of the imaginary part of the DF, Fig. 5.2, we distinguish between ordinary (bright A and B exciton, dark C exciton) and extraordinary (bright C exciton, dark A and B excitons) light polarization. While for a compressive biaxial strain of $\varepsilon_b = -0.02$, as well as vanishing strain, the A- and B-exciton peak can be found at lower energies than the C-exciton peak, this situation changes with larger tensile strains. For $\varepsilon_b = 0.02$ we find that the ordering of the peaks is interchanged. We explain this behavior via the large strain deformation potential of the Γ_{7-v} band (see preceding section) and, hence, also expect this behavior to occur for compressive uniaxial strain (cf. Fig. 5.1).

The strain dependence of the peaks related to the A/B excitons and the C exciton can be observed in experiments due to the polarization dependence (cf. Fig. 5.2). Measuring the *exchange* of the ordering of the A/B and the C exciton is difficult since the necessary strains are large. We give a more detailed investigation of the corresponding exciton binding energies in Ref. [90]. Furthermore, we found that the strain also influences the peak positions and even the optical anisotropy at higher photon energies. When information about band orderings is derived from measurements, knowledge about possible strain in the sample is inevitable.

5.2 Non-equilibrium wurtzite structure: MgO and CdO

When rs-MgO or rs-CdO are mixed with wz-ZnO they can abandon their equilibrium rs crystal structure and adopt the wz structure under certain (non-equilibrium) conditions. In the Zn-rich regime this has been experimentally observed for both of these oxides [102, 103, 164–166]. Though we apply a thermodynamic approach to study the isostructural and heterostructural *alloys* $Mg_xZn_{1-x}O$ or $Cd_xZn_{1-x}O$ in Chapter 6, we focus in this section on an

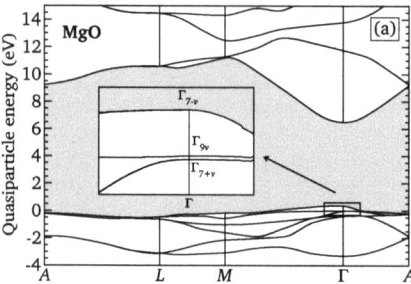 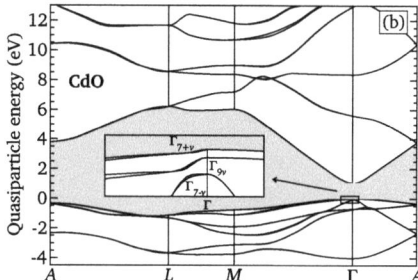

Figure 5.3: Quasiparticle band structure of wz-MgO (a) and wz-CdO (b), including spin-orbit coupling. The valence-band maximum has been used as energy zero and the fundamental gap region is shaded. The insets schematically show the band ordering at the top of the valence bands.

ab-initio prediction of the electronic band structure as well as the properties of the optical absorption edge for pure MgO and CdO in the wz crystal structure. Information about bulk wz-MgO or wz-CdO crystals is experimentally hardly accessible, since no bulk samples are available for the non-equilibrium wz polymorphs. However, knowledge of e.g. fundamental band gaps, CF or spin-orbit splittings, and the lowest optical transitions can be helpful in understanding the aforementioned mixtures of MgO or CdO and ZnO.

Therefore, in this section we employ atomic coordinates that we derived from total-energy minimizations for the wz structure of MgO and CdO within DFT-GGA before [19, 23]. Subsequently, we calculate the corresponding QP band structures within the HSE+G_0W_0 approach and also include SOC (cf. Section 3.1).

5.2.1 Quasiparticle energies

Band structures including spin-orbit coupling

For both systems, wz-MgO and wz-CdO, we plot the QP band structures including SOC in Fig. 5.3. In the case of wz-MgO we find a direct fundamental band gap of 6.52 eV at the Γ point. Besides this, the inset in Fig. 5.3(a) indicates that the VB ordering in wz-MgO is $\Gamma_{7-v}-\Gamma_{9v}-\Gamma_{7+v}$ which differs from the one we observed for wz-ZnO (cf. Section 4.1.2). Furthermore, the Γ_{7-v}-derived band anti-crosses the other two VBs along the direction parallel to the c axis. Contrary to this, wz-CdO [cf. Fig. 5.3(b)] shows the same band ordering as wz-ZnO and the Γ_{7+v}-derived band is very dispersive in the xy plane, i.e., perpendicular to the c axis. As a consequence it anti-crosses the Γ_{9v} and Γ_{7-v} VBs in the direct vicinity of Γ in this plane [cf. inset of Fig. 5.3(b)] and the characters of the bands change accordingly. We attribute the

5.2 Non-equilibrium wurtzite structure: MgO and CdO

		wz-MgO	wz-CdO
without SOC	$E_g = \varepsilon^{QP}(\Gamma_{1c}) - \varepsilon^{QP}(\Gamma_{5v})$	6.52	1.06
	$\Delta_1^{no\ SOC} = \varepsilon^{QP}(\Gamma_{5v}) - \varepsilon^{QP}(\Gamma_{1v})$	−373.9	76.8
with SOC	$\varepsilon^{QP}(\Gamma_{9v}) - \varepsilon^{QP}(\Gamma_{7+v})$	26.1	−23.3
	$\varepsilon^{QP}(\Gamma_{9v}) - \varepsilon^{QP}(\Gamma_{7-v})$	−357.8	65.2
	Δ_1^{qc}	−369.6	73.1
	$\Delta_2^{qc} = \Delta_3^{qc}$	12.6	−10.4

Table 5.2: Electronic structure around the fundamental gap of wz-MgO and wz-CdO: Gap E_g (in eV), valence-band splittings $\varepsilon^{QP}(\Gamma_{9v}) - \varepsilon^{QP}(\Gamma_{7+v})$ and $\varepsilon^{QP}(\Gamma_{9v}) - \varepsilon^{QP}(\Gamma_{7-v})$ (in meV) as well as the derived quantities Δ_1, Δ_2, Δ_3 (in meV).

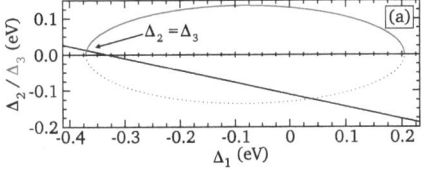
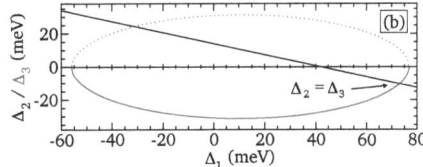

Figure 5.4: Spin-orbit-splitting constants Δ_2 (black) and Δ_3 (gray) for wz-MgO (a) and wz-CdO (b) as a function of the Δ_1 constant that is related to the crystal-field splitting. All quantities are given in eV (meV) and the quasi-cubic approximation for negative (positive) Δ_1 is indicated for wz-MgO (wz-CdO).

two different VB orderings that we observe for wz-MgO (no d electrons) and wz-ZnO/wz-CdO (containing d electrons) to the hybridization of p and d states at the VBM, which is in wz crystals, in contrast to the rs case (cf. Section 4.1.2), symmetry-allowed at Γ. Consequently, we also ascribe the direct fundamental band gap occurring at the Γ point to this effect. This gap is with 1.06 eV significantly smaller than the direct Γ gap of rs-CdO, whereas we found rs-CdO to be an indirect semiconductor (cf. Section 4.1.1).

In addition, we employ the **k·p** theory for the wz crystal structure [119] to derive the splittings Δ_1, Δ_2, and Δ_3 from the QP energies. The possible solutions are plotted for the non-equilibrium polymorphs wz-CdO and wz-MgO in Fig. 5.4 and are given for the quasi-cubic approximation ($\Delta_2^{qc} = \Delta_3^{qc}$) in Table 5.2. For the CF splitting of the uppermost VBs without SOC we obtain $\Delta_1^{no\ SOC} = -374$ meV for MgO and $\Delta_1^{no\ SOC} = 77$ meV for CdO. The absolute values of the SOC-related constants $\Delta_2^{qc} = \Delta_3^{qc}$ almost agree for wz-MgO and wz-CdO, whereas their signs differ (cf. Table 5.2). They amount to 13 % of the CF splitting for wz-CdO and only 3.4 % for wz-MgO. Comparison to the respective results for wz-ZnO (cf. Table 4.2) reveals the chemical trend of decreasing values for $\Delta_2^{qc} = \Delta_3^{qc}$ along the row MgO, ZnO, CdO. The different VB ordering causes the sign change of all three splittings when going from wz-MgO to wz-ZnO or wz-CdO and has been traced back to the influence of the pd hybridization.

	wz-MgO	wz-CdO
$m_M^*(\Gamma_{7c})$	0.36	0.22
$m_K^*(\Gamma_{7c})$	0.44	0.33
$m_A^*(\Gamma_{7c})$	0.34	0.18
$m_M^*(\Gamma_{7+v})$	0.59	0.26
$m_K^*(\Gamma_{7+v})$	1.60	0.60
$m_A^*(\Gamma_{7+v})$	20.05	2.55
$m_M^*(\Gamma_{9v})$	10.92	2.43
$m_K^*(\Gamma_{9v})$	5.34	2.19
$m_A^*(\Gamma_{9v})$	6.47	2.59
$m_M^*(\Gamma_{7-v})$	4.28	2.24
$m_K^*(\Gamma_{7-v})$	4.38	2.27
$m_A^*(\Gamma_{7-v})$	0.37	0.20

Table 5.3: Effective masses m^* (in units of the free-electron mass m) at the Brillouin zone center along the $\Gamma-M$, $\Gamma-K$, and $\Gamma-A$ directions for wz-MgO and wz-CdO. Values are given for the lowest conduction band and the three uppermost valence bands.

Effective masses

Parabolic fits to the QP band structures (including SOC) in the close vicinity of Γ allow the derivation of the effective masses for the lowest CB and the uppermost three VBs. These values (cf. Table 5.3) confirm the aforementioned anisotropic behavior, as well as the crossing and anti-crossing of the bands, with regard to different directions in the BZ. While the lowest CB and the Γ_{9v}-associated VB are comparably isotropic, with effective masses not differing by more than 50%, we find a difference of more than one order of magnitude for the masses of the Γ_{7+v}- and Γ_{7-v}-associated VBs for the three high-symmetry directions $\Gamma-M$, $\Gamma-K$, and $\Gamma-A$. Consequently, the Γ_{7+v}-associated band is the light-hole band in K and M directions and the heavy-hole band in the A direction, whereas the opposite is true for the Γ_{7-v}-associated band. Again, the anisotropy of the in-plane effective masses m_M^* and m_K^* indicates that the corresponding bands are not completely parabolic within the **k**-space region used for the fitting (cf. Section 4.1.2, page 54).

5.2.2 Optical properties of the absorption edge

Instead of the computationally expensive solution of the BSE we chose to study only the matrix elements of the momentum operator for the non-equilibrium wz polymorphs of MgO and CdO. Their values are calculated (using the longitudinal approximation [83]) from the HSE wave functions (without spin-orbit coupling) and are given in Table 5.4. In these values the polarization anisotropy that arises from the dipole selection rules for the respective transitions in the hexagonal crystal structure is reflected. In Table 5.4 only the matrix elements for the $\Gamma_{1v} \rightarrow \Gamma_{1c}$ transition [cf. Fig. 4.4(a)], which is allowed for extraordinary light polarization, and the $\Gamma_{5v} \rightarrow \Gamma_{1c}$ transition, allowed for ordinary polarization, are shown. In contrast

5.2 Non-equilibrium wurtzite structure: MgO and CdO

	wz-MgO	wz-CdO
$\lvert p_{x/y}\rvert^2\,(\Gamma_{5v}\to\Gamma_{1c})$	0.191	0.099
$\lvert p_z\rvert^2\,(\Gamma_{1v}\to\Gamma_{1c})$	0.186	0.103
$E_{\mathrm{B}}(A)$	535	15
$E_{\mathrm{B}}(B)$	435	11
$E_{\mathrm{B}}(C)$	402	11

Table 5.4: Squares $\lvert p\rvert^2$ of (allowed) matrix elements of the momentum operator (in $\hbar^2/a_{\mathrm{B}}^2$) perpendicular ($\lvert p_{x/y}\rvert^2$) and parallel ($\lvert p_z\rvert^2$) to the c axis. The exciton binding energies E_{B} (in meV) are calculated using purely electronic screening.

to wz-MgO, the states at the VBM of wz-ZnO or wz-CdO show a significant contribution of d-type wave functions. Since transitions from d-related states into the s-like CBM (at Γ) are dipole-forbidden, the corresponding matrix elements are smaller when the involved states show some d character. In the case of wz-ZnO this reduction is not as strong [91] since the cation-anion bond length is smaller compared to the other two oxides [19], even though a certain d contribution to the VBM occurs. For all three oxides $\lvert p_{x/y}\rvert^2$ (perpendicular polarization) is almost equal to $\lvert p_z\rvert^2$ (parallel polarization), which we attribute to the similarity of the corresponding bond lengths within one material.

Using a four-band $\mathbf{k}\cdot\mathbf{p}$ model [167] generalized for the wz structure [168] we relate the matrix elements of the momentum operator to effective electron masses by means of the expression

$$m^*(\Gamma_{7c}) = \frac{m}{1+E_p/E_g}, \quad \text{with } E_p \approx \frac{2}{m}\lvert p_{x/y}\rvert^2 \approx \frac{2}{m}\lvert p_z\rvert^2. \tag{5.2}$$

With the matrix elements from Table 5.4 and the gaps of the wz polymorphs (cf. Table 5.2) we find $m^*(\Gamma_{7c}) = 0.39\,m$ for wz-MgO and $m^*(\Gamma_{7c}) = 0.16\,m$ for wz-CdO, which is in good agreement with the average of the corresponding inverse masses (cf. Table 5.3).

An approximate description of exciton-binding energies arising from a two-band Wannier-Mott model [61, 158] leads to a hydrogen-like series given by

$$E_{\mathrm{B}} = R_\infty \cdot \frac{\mu}{m\,\varepsilon_{\mathrm{eff}}^2}\frac{1}{n^2}, \tag{5.3}$$

where R_∞ is the Rydberg constant. For the reduced electron-hole mass μ that enters the model we employ the average of the inverse masses along the different directions in $\mathbf{k}$ space (cf. Table 5.3). The parabola-like shape of the lowest CB and the uppermost three VBs of wz-MgO and wz-CdO (see Fig. 5.3) allows this approximation which yields the binding energies for the A, B, and C excitons [cf. Fig. 4.12(b)] as the $n=1$ states for each of the respective band pairs via Eq. (5.3). Within this two-band model the screening of the electron-hole interaction is described via a static dielectric constant, which we approximate by the values derived within the IPA [23], i.e., $\varepsilon_{\mathrm{eff}} = 3.02$ (wz-MgO) and $\varepsilon_{\mathrm{eff}} = 13.74$ (wz-CdO). However, the quadratic dependence of Eq. (5.3) on the screening constant points out how sensitive the exciton-binding energies are to the description of the screening. Of course, the exciton-binding energies (given in Table 5.4) calculated from Eq. (5.3) do not take interactions between the Γ_{9v}-, Γ_{7+v}-, or Γ_{7-v}-related VBs into account. On the other hand, the remarkable decrease of E_B when going from MgO to CdO is clearly related to the large difference of the respective dielectric constants ε_{eff} and is therefore true despite the approximative calculation of E_B.

5.3 Summary

Two selected distortions of the ideal crystal structure of the three group-II oxides MgO, ZnO, and CdO were studied in this chapter. The impact of uniaxial or biaxial strain on the QP energies at the Γ point was investigated for wz-ZnO. We found that only compressive uniaxial strain or tensile biaxial strain of about 2 % can turn the CF split-off Γ_{7-v} level into the uppermost VB, whereas the spin-orbit-related splitting only slightly changes its value as strain varies. Due to the polarization dependence of the corresponding optical transition-matrix elements, the respective shifts of the excitonic peaks should be distinctively visible in optical measurements. In addition, we calculated the QP band structure, including SOC, for the non-equilibrium polymorphs wz-MgO and wz-CdO and used these results to derive formerly unknown $\mathbf{k} \cdot \mathbf{p}$ parameters, effective masses, and the optical properties around the absorption onset.

6 Pseudobinary alloys: Isostructural versus heterostructural MgZnO and CdZnO

> Denn auf Mischung kommt es an.
>
> Johann Wolfgang von Goethe
> Faust II

In the preceding chapter different strains as well as the crystal structure were proven to influence the electronic and the optical properties of the group-II oxides by modifying, for instance, the fundamental band gap or the band ordering. When exploiting such deviations from the equilibrium structure in order to *design* certain properties of group-II oxide compounds, a possible method is to alloy ZnO with MgO or CdO. Oftentimes it is desirable to control the fundamental band gap for designated applications and devices that are associated with optoelectronics. It has been observed experimentally [100, 101, 169] that the absorption onset can be tuned, e.g. from about 3.4 eV (*wz*-ZnO) up to $\approx$ 4.4 eV ($Mg_xZn_{1-x}O$), which corresponds to the ultraviolet spectral region. Conversely, pseudobinary $Cd_xZn_{1-x}O$ alloys feature smaller gaps that render them suitable for devices operating in the visible spectral range [102].

Unfortunately, isostructural combinations of ZnO and MgO or CdO seem to be thermodynamically unstable because their mixing enthalpy in either the *rs* structure or the *wz* structure is positive [170]. On the other hand, their heterostructural alloys appear to be stable under certain conditions [102, 103, 164–166]. From a theoretical point of view we expect a change of the atomic coordination from fourfold (*wz*) to sixfold (*rs*) with increasing Mg or Cd content which is, in turn, reflected in alloy properties that are very sensitive to the various techniques used for the sample preparation. Therefore, we do not only investigate the alloys under thermodynamic equilibrium conditions by studying their mixing free energy, but, in addition, also take non-equilibrium situations into account. Knowing the atomic geometry of the alloys is essential to calculating their electronic structure and optical properties.

6.1 Thermodynamic properties and lattice structure

The basis of our investigation of alloys is a cluster expansion that relies on 16-atom clusters for the *rs* as well as the *wz* crystal structure (see Section 2.5.1 and Appendix A.1). We start with a total-energy minimization within DFT-GGA to obtain the equilibrium lattice geometry (including fully relaxed atomic coordinates) along with the total energy for one representative of each cluster class. The temperature- and composition-dependent properties of the macroscopic alloys are calculated via the Connolly-Williams method [72, 73], Eq. (2.85), hence the cluster fractions x_j must be determined for x and T. By employing different approaches for computing the x_j, we account for thermodynamic equilibrium and non-equilibrium conditions as they emerge from various experimental techniques, temperatures, and substrate types. In the literature the fabrication of $Mg_xZn_{1-x}O$ films using a variety of methods has been reported: pulsed-laser deposition (PLD) at growth temperatures of 950…1050 K [169], radio-frequency-magnetron sputtering at 700 K [171], and reactive-electron-beam evaporation (REBE) at a substrate temperature of 550 K [172]. $Cd_xZn_{1-x}O$ layers have been prepared by molecular-beam epitaxy with a growth temperature as low as 450 K [102], (plasma-enhanced) metal-organic chemical vapor deposition (MOCVD) at 625 K [103, 173], and PLD at 700 K [174].

The Gibbs free energy is the thermodynamic potential which describes the equilibrium of a system for a fixed temperature and pressure. In this work, the thermodynamic equilibrium conditions are accounted for by cluster fractions x_j that are calculated within the GQCA, i.e., under the constraint of a minimal *Helmholtz (mixing) free energy* (the difference to the Gibbs free energy vanishes for solids at low pressures). In addition, we include two non-equilibrium situations by employing a SRS model as well as a MDM (cf. Section 2.5 and Ref. [76]). The influence of the temperature is studied via the temperature-dependent mixing entropy for (i) room temperature ($T = 300$ K) and (ii) an exemplary growth temperature of $T = 1100$ K.

6.1.1 Mixing free energy

GQCA for isostructural and heterostructural alloys

We investigate the two isostructural *rs* and *wz* alloys and compare them to the heterostructural system, where we take the clusters for both crystal structures into account. Therefore, the index j in the equations in Section 2.5 runs to $J = 21$ (*wz* clusters only), $J = 15$ (*rs* clusters only), or $J = 37$ (both types of clusters) accordingly (see Table A.1). The minimization of ΔF for given x and T has to be performed independently for each situation. The energies of the respective *equilibrium* crystal structures (*rs*-MgO, *wz*-ZnO, *rs*-CdO) are used as levels of

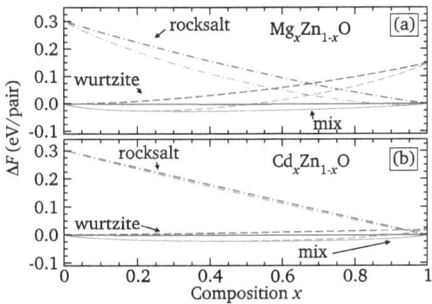

Figure 6.1: Mixing free energy $\Delta F(x,T)$ of $Mg_xZn_{1-x}O$ (a) and $Cd_xZn_{1-x}O$ (b) alloys versus composition x. The curves are obtained using the GQCA with wurtzite clusters only (dashed), rocksalt clusters only (dash-dotted), or both types of clusters (solid). In all cases results are shown for $T = 300$ K (dark gray curves) and $T = 1100$ K (light gray curves). The equilibrium crystal structures wz ($x = 0$) and rs ($x = 1$) have been used as energy zero.

reference.

For $Mg_xZn_{1-x}O$ the mixing free-energy curves for the wz alloys and the rs alloys in Fig. 6.1(a) intersect at $x \approx 0.67$, independent of the temperature. We interpret this as a tendency for a transition from preferred fourfold coordination (wz) to preferred sixfold coordination (rs) at that composition under equilibrium conditions. In addition, the difference of the mixing free energy per cluster of the heterostructural alloys and that of the respective isostructural cases exceeds 25 meV, i.e. k_BT at room temperature, for approximately $0.10 \leq x \leq 0.98$ ($0.28 \leq x \leq 0.93$) for $T = 300$ K ($T = 1100$ K). Hence, for these values of x both, rs as well as wz clusters, significantly contribute to the alloy material. These tendencies agree well with results of T. Minemoto et al. [175] (Z. Vashaei et al. [176]) who report predominantly wz structure below $x \leq 0.46$ ($x \leq 0.34$) and mainly rs structure for $x \geq 0.62$ ($0.65 \leq x \leq 0.97$). X-ray diffraction measurements of thin-film samples by Bundesmann et al. [164] revealed the hexagonal wz structure for $x \leq 0.53$ and the cubic rs structure for $x \geq 0.67$. Films grown by REBE (see Ref. [172]) lead to hexagonal $Mg_xZn_{1-x}O$ up to $x = 0.51$ and to cubic $Mg_xZn_{1-x}O$ above $x = 0.55$. Two theoretical studies [170, 177] report an intersection of the free-energy curves of the isostructural alloys at $x \approx 0.33$ (in contrast to our value of $x \approx 0.67$) because they found almost the same energy difference for rs-ZnO and wz-ZnO as we do for rs-MgO and wz-MgO, and vice versa. Since Fan et al. [177] do not include the d electrons of Zn or Cd in their calculations and Sanati et al. [170] do not comment on if they do or not, we assume that this may contribute to the different behavior. Also, the XC functional (LDA or GGA) seems to affect this issue, as discussed in Ref. [76].

In the case of the $Cd_xZn_{1-x}O$ alloys we study the mixing free energy in Fig. 6.1(b) where we observe a crossing of the curves for the isostructural alloys roughly at a Cd content of $x \approx 0.95$. Moreover, we find that the result from the mixed statistics differs less than 25 meV (per cluster) from the curve for the pure wz structure up to compositions x of about 0.17 (0.59)

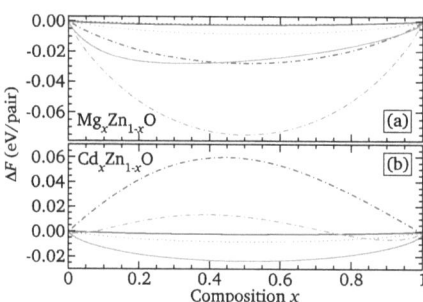

Figure 6.2: Mixing free energy $\Delta F(x, T)$ of $Mg_xZn_{1-x}O$ (a) and $Cd_xZn_{1-x}O$ (b) alloys versus composition x for $T = 300$ K (dark gray) and $T = 1100$ K (light gray). The solid curves are computed using cluster fractions from the GQCA. The dotted curves are obtained for the MDM while the dash-dotted curves are calculated using the ideal cluster fractions. All curves result from the combined statistics with both wurtzite- and rocksalt-type clusters. The respective composition end points have been used as zero (see text).

for $T = 300$ K ($T = 1100$ K). This indicates that, especially for the high-temperature case, a large part of the clusters that form the alloy shows the *wz* structure, which is clearly related to the small energy difference between the *rs*-CdO and the *wz*-CdO phases [19]. This explains why experimental studies of $Cd_xZn_{1-x}O$ give an ambivalent picture: While two groups report very low thermodynamic solubility limits of only $x \approx 0.07$ [174] or phase separation at even lower Cd concentrations [173], another group observed compositions of up to $x = 0.32$ for samples produced by means of a highly non-equilibrium growth mode [102] – unfortunately, they have not tried for higher concentrations. In addition, the *wz* crystal structure has been reported for plasma-enhanced MOCVD layers up to $x = 0.697$ [103], which is confirmed by a transition from *wz* to *rs* structure at $x = 0.7$ of films deposited by MOCVD [178]. Apparently, experiments find the change of the crystal structure at lower Cd concentrations than we predicted from the intersection of the ΔF curves for *wz* and *rs* in Fig. 6.1(b). On the other hand, the high-temperature curve of the heterostructural alloy significantly deviates from that of the isostructural *wz* alloy above Cd contents of about 0.7, which may partly explain the experimental findings [178].

The three different statistics for heterostructural alloys

In Fig. 6.2 we compare the mixing free energies that result from the cluster fractions calculated within the GQCA, the SRS model, and the MDM for the heterostructural alloys. The energy of the respective equilibrium crystal structure has been used as a level of reference at $x = 0$ and $x = 1$, except for the curves obtained using the ideal cluster fractions x_j^0. Since the x_j^0 [cf. Eq. (2.92)] do not depend on the cluster energies but only on the numbers n_j and $n - n_j$ of cations of type A and type B the sum of the x_j^0 for all *wz* clusters and that for all

6.1 Thermodynamic properties and lattice structure

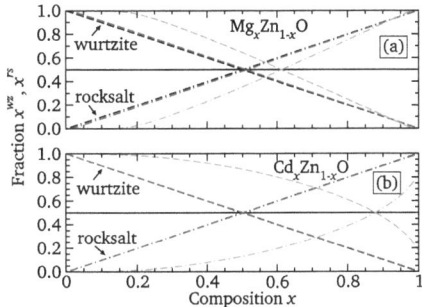

Figure 6.3: Crystal-structure fractions x^{wz} (dashed curves) and x^{rs} (dash-dotted curves) of $Mg_xZn_{1-x}O$ (a) and $Cd_xZn_{1-x}O$ (b) alloys versus composition x. We calculated the curves using the cluster fractions from the GQCA for $T = 300$ K (dark gray curves) and $T = 1100$ K (light gray curves). The solid black lines are obtained using the ideal cluster fractions. For comparison the black curves show the result from the MDM for both crystal structures.

rs clusters give the same total weight of 0.5, independent of the composition x. Therefore, to compare this approach to the other distributions of the x_j in Fig. 6.2, we set the mixing free energies at $x = 0$ and $x = 1$ to zero for each curve resulting from the SRS model. Otherwise, $\Delta F(x = 0, T) > 0$ and $\Delta F(x = 1, T) > 0$ would result for temperatures $T > 0$ K due to the weights x_j^0.

The mixing free energy of the heterostructural $Mg_xZn_{1-x}O$ alloy is lower than zero for all x and T using any of the three approaches for calculating the x_j because the excess energies are negative for all $Mg_{n_j}Zn_{n-n_j}O_n$ clusters [76]. Consequently, all three statistics agree in predicting the heterostructural $Mg_xZn_{1-x}O$ system to be a random alloy without a miscibility gap and as having no tendency for binodal or spinodal decomposition [71]. Contrarily, the ΔF curves for $Cd_xZn_{1-x}O$ resulting from the SRS model [cf. Fig. 6.2(b)] are qualitatively different as they show minima and inflection points with positions that strongly depend on the temperature. As discussed in detail in Refs. [68, 76], this is an indication of a phase transition between random and phase-separated alloys and might explain the low solubility of Cd in ZnO [174], especially when the SRS model describes such growth experiments well. This model predicts a phase transition because it neglects the large energetic differences between the two crystal structures by assigning the ideal weights to the clusters. In addition, Fig. 6.2 points out that the GQCA and the MDM coincide for both materials at low temperatures, i.e., we find the alloys being almost entirely decomposed into the clusters of the binary end components.

6.1.2 Structural composition of heterostructural alloys

Within the description of heterostructural alloys we define the *wz* character x^{wz} and the *rs* character x^{rs} of the system as the sum over the corresponding weights, i.e.,

$$x^{wz} = \sum_{j=0}^{21} x_j \text{ and } x^{rs} = \sum_{j=22}^{37} x_j \qquad (6.1)$$

with $x^{wz} + x^{rs} = 1$ (cf. Table A.1). In Fig. 6.3 we plot these relative contributions of clusters with *wz* or *rs* crystal structure versus the composition x for given temperatures T to derive information about the dominating crystal structure in the alloy, depending on the thermodynamic conditions and the cluster statistics (GQCA, SRS, MDM). Clearly, the SRS model for the cluster fractions gives rise to equal contributions of *rs* and *wz* clusters. We also confirm our previous discussion of the results for the mixing free energy by finding the GQCA curves close to (coinciding with) the MDM results for $Mg_xZn_{1-x}O$ ($Cd_xZn_{1-x}O$) for $T = 300$ K. As expected, higher preparation temperatures tend to move the intersection $x^{wz} = x^{rs}$ to larger Mg or Cd molar fractions x. More specifically, we find that point at about $x = 0.5$ ($T = 300$ K) and $x = 0.6$ ($T = 1100$ K) for $Mg_xZn_{1-x}O$. The temperature dependence is more pronounced for $Cd_xZn_{1-x}O$ and the intersection for $T = 1100$ K occurs at about $x \approx 0.87$. Consequently, the local crystal structure of the $Cd_xZn_{1-x}O$ alloy depends much more on the actual growing conditions which explains the ambivalent experimental findings for that material system [102, 103, 173, 174, 178].

6.2 One-particle excitations

The preceding section elucidated the thermodynamic properties and the structural composition of iso- and heterostructural $Mg_xZn_{1-x}O$ and $Cd_xZn_{1-x}O$ alloys under different thermodynamic conditions. In the following we want to use this knowledge and investigate the electronic structure by means of the HSE+G_0W_0 approach (cf. Section 3.1). This method has been proven in Section 4.1 to provide reliable results for the QP energies of the group-II oxides. In this section we calculate the band structures and DOS for the different clusters and, as before, we include the spin-orbit interaction in a perturbative manner (see Section 3.1.3). Computational parameters are given in Appendix A.2. We would like to mention that the QP energies for the binary MgO, ZnO, or CdO clusters differ from the results discussed in Chapter 4 due to a modification of the VASP-PAW implementation of the GW approximation. As a consequence, we observe slightly smaller band gaps for the group-II oxides (details in Appendix A.2).

6.2 One-particle excitations

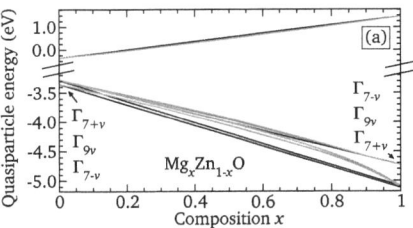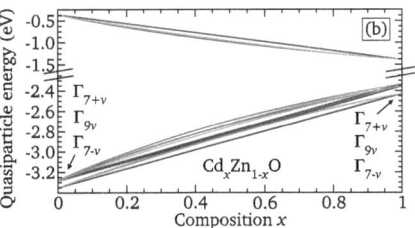

Figure 6.4: Quasiparticle energies (including the spin-orbit interaction) of the uppermost three valence states and the lowest conduction state at the Γ point for wz-$Mg_xZn_{1-x}O$ (a) and wz-$Cd_xZn_{1-x}O$ (b) alloys versus composition x. The branch-point energy (cf. Section 4.1.3) has been used as energy zero. We calculated the curves using the cluster fractions from the GQCA for $T = 300$ K (dark gray curves) and $T = 1100$ K (light gray curves). The medium gray lines are obtained using the ideal cluster fractions. For comparison the black curves that represent the result from the MDM are included. The band ordering is indicated for the binary end components. The similarity of some of the curves is discussed in the text.

6.2.1 Quasiparticle band structures

As discussed in Section 4.1.3, a universal level of reference for the QP energies is necessary when absolute energetic positions of bands are compared for different materials. We face the same problem when comparing the band structures for all the different clusters on an absolute energy scale. For that purpose, we calculate the BPE of each cluster cell and use this value as the *one* universal reference energy for all clusters.

Isostructural alloys: wurtzite

For each cluster cell we extracted the calculated QP energies at the Γ point for the lowest CB and for the uppermost three VBs. While these states were classified as Γ_7 or Γ_9 for the binary end components in Sections 4.1.2 and 5.2, this is no longer possible for the clusters that contain two *different* types of cations. Due to the reduced lattice symmetry the previously introduced definitions for the spin-orbit or the CF splitting do not hold. Interpreting the band structures of alloys in terms that only exist for the binary end components is impossible. For the same reason relating the energy states that occur in different cluster cells is difficult and we merely identify them by their order. However, we cannot access information about possible crossings due to this way of assigning the bands. Using Eq. (2.85) we obtain the configurational averages within the GQCA, the SRS, and the MDM as a function of x and T.

For wz-$Mg_xZn_{1-x}O$ the plot in Fig. 6.4(a) indicates that the GQCA results for $T = 300$ K and $T = 1100$ K almost agree with each other and that the SRS curves match the high-temperature

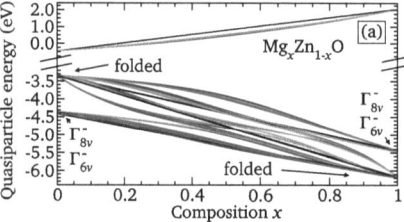 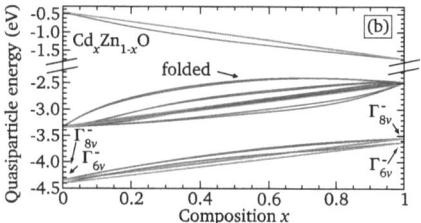

Figure 6.5: Quasiparticle energies (including the spin-orbit interaction) of the uppermost three valence states and the lowest conduction state at the Γ point for rs-$Mg_xZn_{1-x}O$ (a) and rs-$Cd_xZn_{1-x}O$ (b) alloys versus composition x. The branch-point energy (cf. Section 4.1.3) has been used as energy zero. We calculated the curves using the cluster fractions from the GQCA for $T = 300$ K (dark gray curves) and $T = 1100$ K (light day curves). The solid medium gray lines are obtained using the ideal cluster fractions and for comparison the black curves show the results from the MDM. For the binary end components the band ordering is indicated.

GQCA ones, whereas the MDM results deviate. More specifically, the MDM predicts larger splittings of the uppermost two VB states for small values of x. Both the SRS and the GQCA agree in finding an increase of the splitting between these two levels only when $x > 0.7$ which is accompanied by a remarkable bowing. This indicates the significant difference between a linear interpolation of the energy levels of the binary end components and the results from averages which take all clusters into account. In the case of the isostructural wz-$Cd_xZn_{1-x}O$ alloy the MDM curves and that obtained using the GQCA for $T = 300$ K show a good agreement [cf. Fig. 6.4(b)].

Isostructural alloys: rocksalt

For the isostructural rs-$Mg_xZn_{1-x}O$ and rs-$Cd_xZn_{1-x}O$ alloys we present the configurational averages of the QP energies for the upper VB region and the lowest CB in Fig. 6.5. In the band structure of pure rs-CdO (as well as pure rs-ZnO) we observed that, due to the pd repulsion, the fundamental band gap is indirect between the VBM at L and the CBM at Γ (see Section 4.1.1). Since the 16-atom cluster cells are larger than the elementary cells of the rs crystal structure (2 atoms), the corresponding BZs of the cluster cells contain *folded* states, e.g., the uppermost states at L are folded to the Γ point for the alloys. In general, it is not possible to distinguish between states that occur at the Γ point also in the BZ of the 2-atom rs cell and states that are merely folded into the BZ of the alloy. At least for rs-$Cd_xZn_{1-x}O$ these folded bands lie at the top of the VBs and are energetically well-separated from the regular states at Γ for all clusters since the pd repulsion occurs in rs-ZnO as well as rs-CdO [cf. Fig. 6.5(b)]. However, in the case of the rs-$Mg_xZn_{1-x}O$ alloy this separation does not

6.2 One-particle excitations

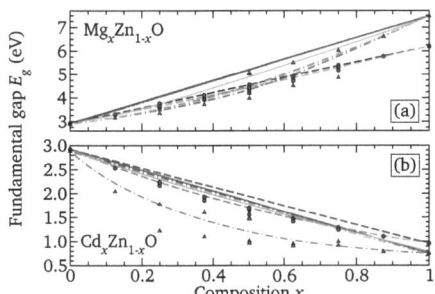

Figure 6.6: Difference of the QP energies of the uppermost valence-band and the lowest conduction-band state for $Mg_xZn_{1-x}O$ (a) and $Cd_xZn_{1-x}O$ (b) alloys versus composition x. The curves are obtained using only wurtzite clusters (dashed lines), only rocksalt clusters (dash-dotted lines), or both type of clusters (solid lines). We use cluster fractions from the GQCA for $T = 300$ K (dark gray curves) and $T = 1100$ K (light gray curves), as well as ideal cluster fractions (medium gray curves). For comparison the black curves show the result from the MDM and the fundamental gaps are also included for the clusters in rocksalt (light gray triangles) and wurtzite (dark gray circles) crystal structure.

occur since pure rs-MgO has a direct gap (due to the lack of pd repulsion). Consequently, the states in the alloy band structure cannot be clearly assigned [cf. Fig. 6.5(a)] and a mixing occurs for intermediate compositions. Due to the difficulties with the assignment of single states, the discussion of integral quantities, such as the DOS or the DFs, is more meaningful for the alloys and will be done in Sections 6.2.2 and 6.3.

Fundamental band gaps

Another quantity which is well-defined for all clusters is the fundamental gap, as the difference of the QP energies of the lowest CB state and the highest VB state. In Fig. 6.6 we show the configurational averages of these values for isostructural as well as heterostructural $Mg_xZn_{1-x}O$ and $Cd_xZn_{1-x}O$ alloys.

For wz-$Mg_xZn_{1-x}O$ we observe a good agreement of the results obtained within the GQCA and the SRS model, whereas these curves differ slightly from the prediction of the MDM [cf. Fig. 6.6(a)]. Since for $Mg_xZn_{1-x}O$ the gap values of the individual rs clusters cover a larger range than those of the wz clusters, both the GQCA as well as the SRS curves deviate more from the MDM results in the rs case. The similarity of the direct gap of wz-ZnO and the indirect gap of rs-ZnO is the reason why the MDM curves of heterostructural $Mg_xZn_{1-x}O$ and of rs-$Mg_xZn_{1-x}O$ are very close. The GQCA for the heterostructural system predicts slightly smaller values. An experimental gap of 4.1 eV has been reported by A. Ohtomo et al. [179] for a Mg content of $x \approx 0.33$ and agrees well with our result for the isostructural alloy as well as with a value of 4.2 eV derived from DFT-LDA in combination with scissors operators for the gaps of wz-MgO and wz-ZnO [180]. Experimental gaps of about 6 eV at $x \approx 0.67$ or 7 eV at $x \approx 0.9$ [181] can be explained by our curves for the heterostructural or the isostructural rs-Mg$_x$Zn$_{1-x}$O alloys. In addition, the occurrence of two different slopes of the curves describing the gap versus the composition for the isostructural rs and wz alloys [cf. Fig. 6.6(a)] agrees with findings reported in Ref. [172].

For Cd$_x$Zn$_{1-x}$O the fundamental gaps we observe for the wz crystal structure are similar to the ones of the rs case since the folded states are the uppermost ones for all rs clusters. Therefore, the results found for the two isostructural alloys are much closer to each other than in the case of Mg$_x$Zn$_{1-x}$O. We can see in Fig. 6.6(b) that the SRS model leads to the largest deviations, while the bowing for the other statistics remains small, indicating a dependence on the cluster statistics and, hence, the sample preparation. When comparing to the composition-dependent gap values reported in Ref. [103] we observe that our calculations tend to slightly underestimate the experimental results. We attribute this to the underestimation of the gap that we already found for pure CdO (see discussion in Section 4.1.1).

In addition, we derive bowing parameters b from the curves in Fig. 6.6 using the expression

$$E_g(x) = (1-x)E_g(A) + xE_g(B) - bx(1-x), \qquad (6.2)$$

where $A =$ ZnO and $B = \{$MgO, CdO$\}$ in the respective crystal structures. For isostructural wz-Mg$_x$Zn$_{1-x}$O we obtain values for b of 0.48 eV (for the SRS as well as the GQCA at $T = 1100$ K) and 0.44 eV (GQCA for $T = 300$ K), which agrees well with a value of 0.56 eV from another calculation [180]. The bowing is with 2.58 eV (SRS), 3.12 eV (GQCA, $T = 300$ K), or 2.74 eV (GQCA, $T = 1100$ K) larger for rs-Mg$_x$Zn$_{1-x}$O, in rough accordance with another calculated result of 3.1 eV [170]. The values for the heterostructural Mg$_x$Zn$_{1-x}$O alloy are strongly temperature-dependent as they vary from 0.07 eV ($T = 300$ K) to 1.30 eV ($T = 1100$ K). For Cd$_x$Zn$_{1-x}$O we find all bowings for room temperature to be smaller than 0.023 eV, whereas the results for $T = 1100$ K are 0.72 eV (wz-Cd$_x$Zn$_{1-x}$O), 0.18 eV (rs-Cd$_x$Zn$_{1-x}$O), or 0.29 eV for the heterostructural alloy. The SRS model yields even larger values of 0.95 eV (wz-Cd$_x$Zn$_{1-x}$O) or 2.51 eV (rs-Cd$_x$Zn$_{1-x}$O). Therefore, for most cases there is a non-linear composition dependence of the fundamental gaps.

6.2.2 Densities of states

By means of HSE+G_0W_0 calculations we obtain the DOS of each cluster in the cluster expansion (see Section 2.5) as well as the individual BPEs, which are used for the absolute energy alignment. Using the cluster fractions from the GQCA, the MDM, and the SRS model, we

6.2 One-particle excitations

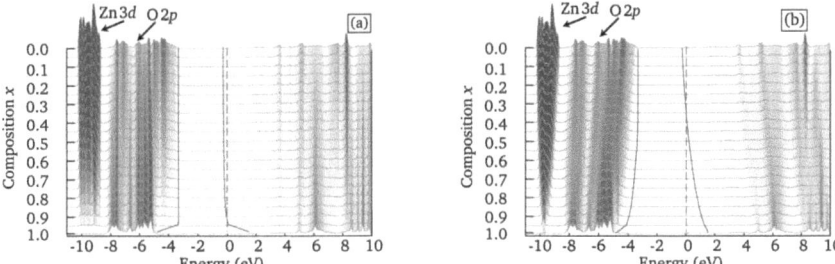

Figure 6.7: Density of states of the isostructural wz-Mg$_x$Zn$_{1-x}$O alloy versus composition x. The branch-point energy (cf. Section 4.1.3) has been used as energy zero (dashed line). We calculated the curves using the cluster fractions from the MDM (a) and the SRS model (b). The black lines indicate where the DOS in the gap region decreases to 0.01 (eV/pair)$^{-1}$.

employ Eq. (2.85) for the values of the DOS at each energy separately in order to access the respective configurational averages for both alloy systems.

Isostructural wz-Mg$_x$Zn$_{1-x}$O alloys

In Fig. 6.7 we compare the results of the DOS for the wz-Mg$_x$Zn$_{1-x}$O alloy, obtained within the MDM and the SRS model, in order to study the differences resulting from the two statistics (the GQCA results strongly resemble those of the SRS model). As expected for the MDM, we find from Fig. 6.7(a) that the DOS curves of pure wz-ZnO linearly transforms into that of the pure wz-MgO as x increases. Consequently, the energy positions and widths of all structures or peaks remain constant and only the respective heights depend on x [see for example the Zn $3d$-related structure in Fig. 6.7(a)]. Contrarily, the SRS model, i.e., when all the different clusters contribute, also captures changes of peak widths or peak positions (relative to the BPE) with varying composition x [cf. Fig. 6.7(b)]. For the two statistics the peak caused by the Zn $3d$ electrons vanishes in different ways with increasing content of Mg in the alloy. In addition, the width of the uppermost VB complex decreases when going from wz-ZnO towards wz-MgO. Figure 6.7 includes lines that indicate where the DOS becomes smaller than 0.01 (eV/pair)$^{-1}$, an indication for the behavior of the VBM and the CBM. For both statistics the trend of these lines points out that at intermediate compositions x, the fraction of clusters with $(n_j/n) < x$ is still significant and leads to a finite DOS.

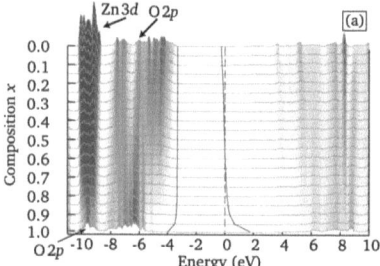 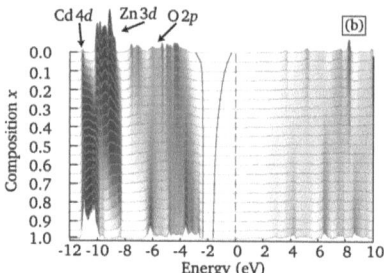

Figure 6.8: Density of states of the heterostructural $Mg_xZn_{1-x}O$ (a) and $Cd_xZn_{1-x}O$ (b) alloys versus composition x, as obtained within the GQCA at $T = 1100$ K. The branch-point energy (cf. Section 4.1.3) has been used as energy zero (dashed line). The black lines indicate where the DOS in the gap region decreases to 0.01 (eV/pair)$^{-1}$.

Heterostructural $Mg_xZn_{1-x}O$ and $Cd_xZn_{1-x}O$ alloys

As one might expect from Fig. 6.3, where we show that the MDM and the GQCA agree very well for room temperature, comparing the DOS for both $Mg_xZn_{1-x}O$ and $Cd_xZn_{1-x}O$ alloys, respectively, shows that the GQCA results for room temperature strongly resemble the ones of the MDM. Since, as discussed above, these plots look merely like a linear transition of the DOS curves of the binary end components, we show in Fig. 6.8 the configurational average calculated within the GQCA for $T = 1100$ K instead. In this case, the GQCA predicts that more than 50 % of the clusters will occur in the wz crystal structure also above $x = 0.5$ (cf. Section 6.1.2).

Interestingly, Fig. 6.8 shows for $Mg_xZn_{1-x}O$ that the peak related to the Zn 3d electrons evolves into an O 2p-derived complex. Above $x \approx 0.7$ a structure related to these O 2p electrons emerges within the Zn 3d states [cf. Fig. 6.8(a)] since the energy position of the VBM with respect to the BPE simultaneously decreases with an increasing x. Both the increase of the fundamental gap as well as the reduced pd repulsion with increasing Mg content in the alloy lead to this decrease of the energy position of the uppermost O 2p derived VB complex with respect to the BPE. Figure 6.8(b) clearly points out for $Cd_xZn_{1-x}O$ how the position of the d-derived states decreases in energy since the Cd 4d electrons are bound stronger than the Zn 3d electrons. Besides this, the O 2p complex appears at higher energies with increasing x.

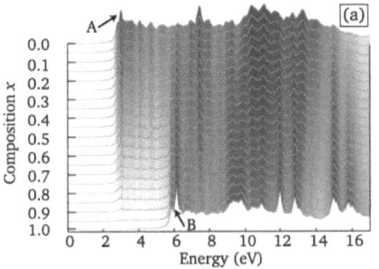
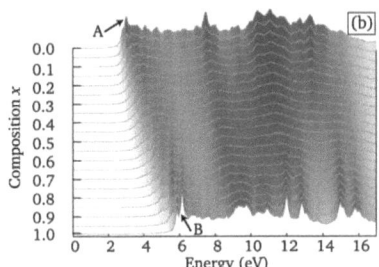

Figure 6.9: Imaginary part of the dielectric function of wz-$Mg_xZn_{1-x}O$ versus composition x calculated using the MDM (a) and the SRS model (b). A (B) labels the excitonic bound state at the absorption edge of wz-ZnO (wz-MgO).

6.3 Dielectric function of wz-$Mg_xZn_{1-x}O$

The optical properties of the alloys, including excitonic and local-field effects, are studied by calculating the imaginary part of the DF. We restrict ourselves to considering only wz-$Mg_xZn_{1-x}O$ due to the tremendous computational effort related to setting up the excitonic Hamiltonian and the subsequent calculation of the DF. Furthermore, this alloy system seems to be well-investigated experimentally and we found, in Section 6.1, that at least for small x the wz crystal structure is of importance. To perform the configurational average for the imaginary part of the DF also at photon energies above the absorption edge we solve one BSE for the low-energy and one for the high-energy part (cf. Sections 3.2 and 4.2) for all 22 clusters of the cluster expansion for wz-$Mg_xZn_{1-x}O$ (see Appendix A.1). The scissors operator Δ of the GGA+U+Δ approach to calculate the underlying electronic structure is also determined for each cluster of the expansion. As done for the DOS, we employ Eq. (2.85) for the values of the imaginary part of the DF at each energy separately so as to obtain the configurational averages. Computational parameters are given in Appendix A.2.

To emphasize the qualitative difference between the results from two different statistics we compare the DF of wz-$Mg_xZn_{1-x}O$, as obtained by means of the MDM, to the results of the SRS model in Fig. 6.9. We find peaks associated with a bound excitonic state at the absorption edge (see discussion in Section 4.2) of pure wz-ZnO (wz-MgO) and indicate them by A (B) in this figure. As before, we observe that the MDM corresponds to a linear transition of the curve for wz-ZnO into the curve for wz-MgO as the value of x increases. Consequently, *both* excitonic peaks (A and B) occur for intermediate values of x in Fig. 6.9(a). Comparing the evolution of peaks A and B as x increases in the plot resulting from the SRS model points out the strong dependence on the cluster statistics. When the DFs of all clusters are taken into account for the configurational average, Fig. 6.9(b) shows that the peaks A and B only occur in the x range close to the binary end components. For intermediate compositions a very broad structure dominates the absorption edge. Such a behavior has been observed in a study of the photocurrents for wz-Mg$_x$Zn$_{1-x}$O for various x [182]. In any case, we observe a strong dependence of the structure of the absorption edge on the cluster fractions. Therefore, in experiment it should be visible if in a real sample larger regions of the pure material occur. Figure 6.9 also depicts the evolution of the peak structures with the composition at higher photon energies and we again find that the SRS model tends to yield broad structures instead of distinct peaks. The origin of these peaks has been discussed for the binary end components in Section 4.2.2.

6.4 Summary

In this chapter we studied isostructural as well as heterostructural pseudobinary Mg$_x$Zn$_{1-x}$O and Cd$_x$Zn$_{1-x}$O alloys by means of a cluster expansion method. We employed different approaches for the calculation of the cluster fractions x_j. Our results for the mixing free energies have been used to understand different experimental findings for composition ranges in which either the wz or the rs crystal structure dominates the alloy.

Using the different cluster statistics and QP energies calculated within the HSE+G_0W_0 approach, we studied the electronic structure of the different alloys (including SOC). We found remarkable bowings for the fundamental band gaps in agreement with experimental findings and other calculated results. The lower symmetry of the alloys' lattice structures renders an interpretation of the evolution of individual states with the cluster composition x infeasible. Moreover, we derived configurational averages for integral quantities, such as the DOS.

Our results for the DF, including excitonic effects, are particularly interesting. Depending on the cluster statistics and, therefore, the preparation conditions of the alloy, we found different trends of the peaks related to bound excitonic states at the absorption edge. Since this behavior should be distinctively observable in experiments, it could contribute to investigations of the alloy's constitution.

7 A point defect: The oxygen vacancy as F-center in *rs*-MgO

> Find out the cause of this effect,
> Or rather say, the cause of this defect,
> For this effect defective comes by cause.
>
> William Shakespeare
Hamlet

The variety of *defects*, i.e., deviations from the ideal atomic pattern of a bulk crystal that can occur in a real crystal, is extensive. As elucidated for strained *wz*-ZnO in Chapter 5 and pseudobinary alloys in Chapter 6, deviations from the ideal crystal structure have an impact on the electronic structure and the optical properties. Defects are categorized with respect to their spatial extension in the crystal. In particular, a *point defect* is not extended in space and is typically restricted to only one or very few atoms, nevertheless, there are still many possibilities for point defects. More specifically, anion vacancies in an ionic crystal can introduce a defect level within the fundamental band gap. As a consequence, the material absorbs light in a narrow spectral region around the wavelength that corresponds to the energetic position of the defect level. When the absorption occurs in the visible spectral range the normally transparent material shows a characteristic color, which is the reason why this type of defect is also called *color center* or *F-center*.

Along these lines, the oxygen vacancy in *rs*-MgO, a very prototypical example for an F-center, has attracted attention for more than five decades [183–186]. In experimentation, two techniques of creating these vacancies are applied: (i) irradiation of high-energetic particles or X-rays, and (ii) the so-called thermochemical reduction, also referred to as additive coloration, where MgO crystals are heated in Mg vapor, i.e., under non-stoichiometric, Mg-rich conditions [183].

The studies related to the oxygen vacancy in *rs*-MgO are part of a collaboration with the groups of Chris G. Van de Walle and Matthias Scheffler. Within this work, we focus on the influence of excitonic effects on the optical absorption from the defect level and use several

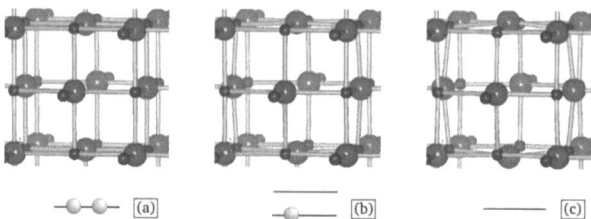

Figure 7.1: Lattice structure of *rs*-MgO in the presence of an oxygen vacancy. The F-center, as the neutral charge state with a filled defect level, is shown in (a). In addition, (b) contains the F^+-center with one electron less, and (c) the F^{2+}-center, where both electrons are removed from the system. (Courtesy of P. Rinke.)

(yet unpublished) results, especially from Patrick Rinke, that constitute important input for our calculations.

7.1 Atomic geometries and charge states

When one oxygen atom is removed from MgO and the oxygen vacancy is created, the entire crystal remains electrostatically neutral. In this case the vacancy level is filled with two electrons, as depicted in Fig. 7.1(a) for the F-center. At the same time, the defect state is the highest occupied energy state in the system and taking out one electron, e.g. by means of an excitation process, turns the F-center into an F^+-center. As a consequence, the defect level splits into two states, one of them being occupied and the other one empty, which renders a spin-polarized description of the problem necessary. In Fig. 7.1(b) we observe that the removal of one electron is accompanied by an outward relaxation of the six next-nearest Mg neighbors of the vacancy. Removing also the second electron from the defect level transforms the F^+-center into an F^{2+}-center, which triggers a further outward relaxation of the next-nearest Mg neighbors, as indicated in Fig. 7.1(c). This behavior of the Mg atoms can be explained by a decrease of the electronic charge density at the vacancy site with the decreasing population of the defect state. Lacking the electron density as a mediator of their positive charges, the Mg ions tend to repel each other. The atomic geometries for the different charge states of the vacancy have been obtained from total-energy minimizations within DFT-LDA.

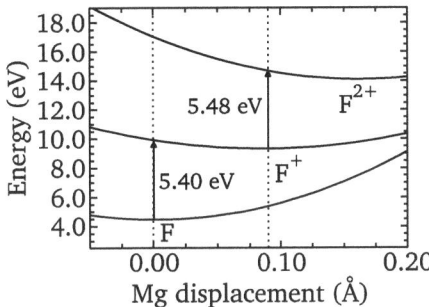

Figure 7.2: Configuration diagram for the different charge states of the oxygen vacancy in rs-MgO. The G_0W_0-corrected total energy is shown for the F-center, the F^+-center, and the F^{2+}-center versus the displacement of the next-nearest Mg neighbors of the vacancy. The arrows indicate the electron removal energy (negative electron addition energy) for the respective equilibrium atomic geometries, which are represented by dotted lines. (Courtesy of P. Rinke.)

7.2 Transition energies and absorption

The optical absorption of the F-center in MgO has been the subject of experimental studies [183, 184]. It has been especially puzzling that the optical absorption spectra of the F- and the F^+-center seem to be very similar, with peaks that strongly resemble each other regarding both their shape and their position in energy [183]. In the literature, the absorption peak of the F-center is reported to occur at 5.03 eV and that of the F^+-center is located at 4.93 eV [184]. Converting F into F^+ by irradiation of light comes along with remarkable lattice relaxations, as indicated in Fig. 7.1.

Studying the corresponding absorption energies and their dependence on the lattice geometry leads to the configuration diagram given in Fig. 7.2. The values shown in this plot result from a newly developed scheme for treating defects [17, 187]. Whereas lattice contributions to the defect formation energies are obtained within DFT-LDA, the removal or addition of an electron is accounted for by energies calculated using a modified PBE0 approach [188] that exactly reproduces the QP gap of an OEPx+G_0W_0 calculation [49, 189] for the bulk material. Furthermore, a recent correction scheme has been applied in order to account for the artificial electrostatic interaction of charged defects in supercell calculations [190]. These state-of-the-art electronic-structure calculations were performed by Patrick Rinke for 63-atom supercells.

As it turns out, modeling the optical excitation by a charged F-center and an uncorrelated electron in the conduction band, as done in Fig. 7.2, is not sufficient. As indicated in Fig. 7.2, the peak of the F-center appears at around 5.40 eV in the calculations, and that of the F^+-center at about 5.48 eV. A comparison with the measured values shows that not only the absolute energy positions are strongly overestimated, but also the energetic ordering of the two is inverted (cf. Fig. 7.4). Since these experiments are based on *optical* excitations of

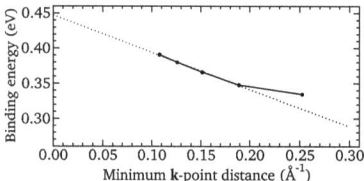

Figure 7.3: Exciton binding energy for the F-center versus the minimum **k**-point distance for different **k** meshes. Extrapolation via a linear fit gives the converged result.

the system, it is of highest interest to explore the influence of the electron-hole interaction. However, excitonic effects have not been taken into account in the plot in Fig. 7.2. In the following section we calculate the exciton binding energies for the transition from the vacancy level into the CBM for the F- and the F^+-center.

7.3 Exciton binding energies

We employ the atomic geometries of the 63-atom cells provided by Patrick Rinke for the two charge states of the vacancy, along with the static electronic dielectric constants determined for both within the IPA, to solve the BSE (cf. Sections 2.4 and 3.2). These screening constants are found to be about 15 % larger than for the ideal bulk *rs*-MgO. In Ref. [61] it has been pointed out how difficult it is to converge the binding energies of Wannier-Mott-like excitons with respect to the **k**-point sampling of the BZ due to the parabolicity of the bands. Although the defect level shows almost no dispersion, the CB is parabolic, hence, this problem persists. Calculations involving defects become particularly difficult due to the large numbers of atoms. In Fig. 7.3 we employ the procedure described in Ref. [61] to obtain converged results for the exciton binding energy by calculating them as a function of the **k**-point sampling and extrapolate linearly to vanishing **k**-point distances. As mentioned before, the situation is even more complicated in the case of the F^+-center since the description of the problem must take spin polarization into account. The solution of the BSE with spin polarization only recently became possible [57]. However, the computational cost increases even more due to the doubled number of bands which must be taken into account. Overall, these calculations for the oxygen vacancy in *rs*-MgO are at the edge of what is possible from a computational point of view.

Using this procedure, we calculated the exciton binding energies for the transition from the defect level into the CBM. We obtained a value of 0.45 eV for the F-center and 0.56 eV for the F^+-center, which shows for both charge states of the vacancy that the electron-hole interaction remarkably influences the optical absorption properties. Subtracting our results from the transition energies calculated using the modified PBE0 approach as discussed earlier,

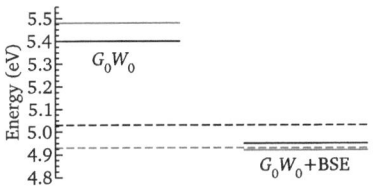

Figure 7.4: Excitation energies of the F-center (black) and the F^+-center (gray). The values obtained from the G_0W_0-corrected scheme are compared to results that include the electron-hole interaction and to experimental findings (dashed lines).

we find that the absorption peak of the F-center occurs at 4.95 eV and that of the F^+-center at 4.92 eV. The plot of these results in Fig. 7.4 shows a good agreement with the experimental observations regarding the relative energetic ordering and the absolute energetic position.

7.4 Summary

In this chapter we were able to show a significant impact of the electron-hole interaction on the energetic position of the defect-related peak in the optical absorption spectrum for the F-center and the F^+-center in *rs*-MgO. The respective exciton binding energies were predicted to be very large and are even responsible for a change in the energetic ordering of the peaks corresponding to the two charge states of the vacancy. The 63-atom supercells used for these calculations constitute a challenge to current computational possibilities. Nevertheless, the agreement with the measured values is impressive and shows how modern *ab-initio* calculations can contribute to unraveling and understanding observations from experiments.

8 Heavy n-doping: Wannier-Mott and Mahan excitons in wz-ZnO

> Is it not? Is it not? Breadth of view, my dear Mr. Mac, is one of the essentials of our profession. The interplay of ideas and the oblique uses of knowledge are often of extraordinary interest.
>
> Arthur Conan Doyle
> Sherlock Holmes, The Valley of Fear

As a branch of semiconductor technology, the modern field of *optoelectronics* is expected to grow extensively in the future, driven, for instance, by the next generation of display devices. Nowadays, it comprises the development of modern solar cells or flat-panel LCD displays; both are exemplary applications that certainly benefit from transparent electrodes, i.e., layers of materials that are simultaneously transparent for visible light and conduct electrical current. As already mentioned, the group-II oxides are gaining importance in this context as *transparent conductive oxides* (TCOs). Their large fundamental band gaps (cf. Chapter 4) render them transparent in the visible spectral range, while, by means of n-doping, free electrons can be introduced, making these materials conductive. For practical applications an efficient charge transport is desirable, which requires sufficiently high free-electron concentrations.

Among the TCOs ZnO again plays an important role, hence, it is very well-investigated, especially from an experimental point of view [191]. Via doping with aluminum or indium [98, 99], samples with free-electron concentrations of more than $5 \cdot 10^{20}$ cm^{-3} can be produced. In contrast, nominally undoped bulk or thin-film samples show electron concentrations of about $10^{13} \ldots 10^{17}$ cm^{-3} [191]. The presence of a degenerate electron gas in the CBs of any material is expected to strongly modify the optical properties with respect to the undoped situation. The free electrons occupy the lowest CB states and, in addition, their presence in the material has a significant influence on the screening of the electron-electron and the electron-hole interaction. A deep theoretical understanding is inevitably necessary.

For that reason we study the interplay of excitonic effects and a degenerate electron gas

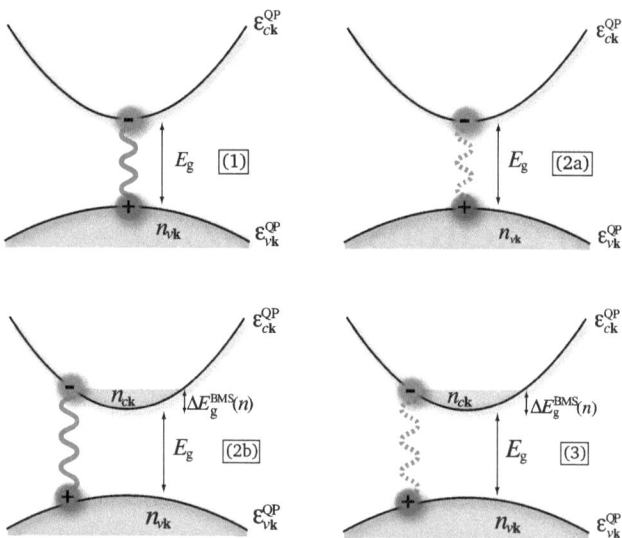

Figure 8.1: Parabolic two-band model for studying the optical absorption of the undoped material (panel 1) and when a free-electron gas is present, i.e., $n_{c\mathbf{k}} \neq 0$. The conduction band $\varepsilon_{c\mathbf{k}}^{QP}$ and the valence band $\varepsilon_{v\mathbf{k}}^{QP}$ are separated by the fundamental gap E_g. We separately investigate the influence of a modified electron-hole interaction (panel 2a) and the Pauli blocking (panel 2b), as well as a combination of both effects (panel 3).

in the lowest CB of wz-ZnO. After gaining insight into the problem by means of a two-band model, we extend the parameter-free BSE approach introduced before (cf. Sections 2.4 and 3.2) to deal with the free electrons. We calculate exciton binding energies, the optical oscillator strengths, and the absorption coefficient. In addition, we discuss the possibility of an excitonic Mott transition. We also explore to which extent inter-conduction-band absorption (ICBA) influences the optical spectrum around the absorption edge.

8.1 Approaching the problem via a two-band model

8.1.1 Effects due to a degenerate electron gas

We begin with a description of the problem within a two-band model, as depicted in Fig. 8.1, to introduce the terms and basic effects related to a degenerate electron gas in the lowest CB of a material. This situation is different from pumping processes and henceforth we assume a fully occupied parabolic VB with the QP energies $\varepsilon_{v\mathbf{k}}^{\mathrm{QP}}$, which is separated by a fundamental gap E_{g} from an empty parabolic CB described by $\varepsilon_{c\mathbf{k}}^{\mathrm{QP}}$. Due to the irradiation of a photon an electron is excited into the CB while leaving a hole in the VB. Both interact with each other via the screened Coulomb interaction, as previously discussed (cf. Section 4.2). The parameters of the model are chosen according to the results for *wz*-ZnO presented in Section 4.1, i.e., we use an effective CB mass of $m_c = 0.3\,m$ and a VB mass of $m_v = 0.5\,m$. The gap is set to $E_{\mathrm{g}} = 3.21$ eV and we employ the same electronic static dielectric constant, $\varepsilon_{\mathrm{eff}} = 4.4$, as for the BSE calculations in Section 4.2. The influence of the degenerate electron gas on the QP gap [see discussion in Section 8.2.1 and Eq. (8.3)] does not influence the physics of the electron-hole interaction.

A first step towards incorporating free electrons in the lowest CB into our description is depicted in panel 2a of Fig. 8.1, where we indicate a modification of the electron-hole interaction. Within the TF model (see Section 2.4.4) one has to replace the screened Coulomb potential of the electron-hole interaction by a Yukawa potential [cf. Eq. (2.73)]. The inverse screening length is equal to q_{TF}, Eq. (2.71), and thereby related to the free-electron density. As a short-range interaction, the Yukawa potential has no bound states when the screening length falls below a certain value, i.e., when the free-electron concentration exceeds a critical value [65, 66]. Such an unbinding of the electron and the hole is the so-called *Mott transition* of the exciton. As pointed out in Ref. [192] different formulae for calculating the respective Mott densities n_{M} of *wz*-ZnO are applied in the literature. Consequently, the results cover a range of several orders of magnitude and there seems to be a fundamental lack of understanding.

Alternatively, we show in panel 2b of Fig. 8.1 that a degenerate electron gas occupies the lowest CB states and, consequently, the respective optical transitions are forbidden due to the Pauli principle. This is referred to as *Pauli blocking* and leads to an enlargement of the optical gap, which is interpreted as a "shift", the so-called Burstein-Moss shift (BMS) [193], $\Delta E_{\mathrm{g}}^{\mathrm{BMS}}(n)$. For a parabolic CB this shift is equal to the Fermi energy of the free electrons [see definition next to Eq. (2.70)]. Obviously, its magnitude depends on the density n_c of the additional electrons. As shown in panel 3 of Fig. 8.1 we ultimately want to investigate the combined effect of the modified electron-hole interaction and the Pauli blocking on the excitons.

8.1.2 Semiconductor Bloch equations

We study the different contributions (cf. Fig. 8.1) using the solution of the semiconductor Bloch equations (see Section 2.4.5 and Ref. [65]) as a computationally feasible approach

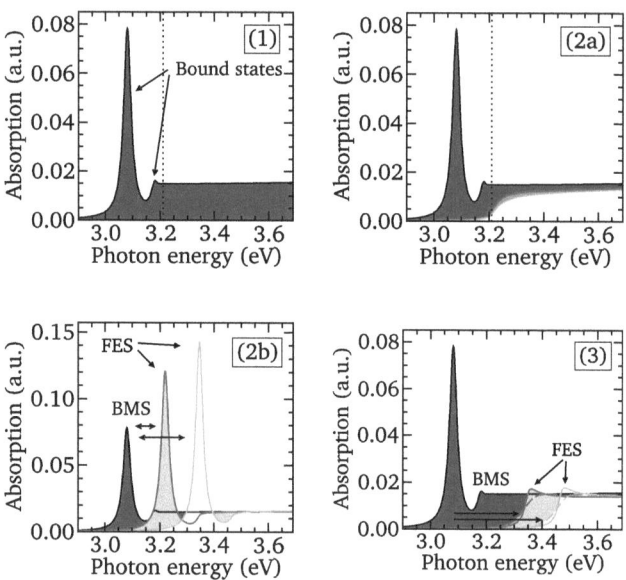

Figure 8.2: Optical absorption at $T = 0$ K versus photon energy calculated within the two-band model for the undoped material (panel 1) and in the presence of a degenerate electron gas. We separately investigate the influence of a modified electron-hole interaction (panel 2a) and the Pauli blocking (panel 2b), as well as a combination of both effects (panel 3). In panel 3 the dashed curves are calculated for $T = 300$ K. We show results for the undoped material (black curves) and for two free-electron densities, i.e., $n_{c,1} = 1.9 \cdot 10^{19}$ cm^{-3} (gray curves) and $n_{c,2} = 4.8 \cdot 10^{19}$ cm^{-3} (light gray curves). The QP gap (without shrinkage) is indicated by a dotted line.

that is well-adapted to the two-band model. The spherically symmetric problem, Eq. (2.80), is discretized and solved via a time-evolution approach [65, 194]. We perform the calculations for two different, experimentally easily achievable free-electron concentrations [97] of $n_{c,1} = 1.9 \cdot 10^{19}$ cm^{-3} and $n_{c,2} = 4.8 \cdot 10^{19}$ cm^{-3}. Panel 1 of Fig. 8.2 also shows the absorption coefficient for the two-band model without additional free electrons. As before (cf. Section 4.2.2), excitonic bound states with large oscillator strength occur below the QP gap.

Modified electron-hole interaction

Accordingly, in panel 2a of Fig. 8.2 we show the optical absorption which results when the electron-hole interaction is modeled by a Yukawa potential. The respective screening lengths for both values $n_{c,1}$ and $n_{c,2}$ are determined as inverse TF wave vectors, Eq. (2.71). We find

8.1 Approaching the problem via a two-band model

that the screening majorly impacts the bound state close to the absorption onset. Already for the two moderately large densities $n_{c,1}$ and $n_{c,2}$ the additional screening is so strong that there is no bound state visible anymore and the Coulomb enhancement at higher photon energies is slightly less pronounced.

Pauli blocking of the lowest conduction-band states

Alternatively, we study the case illustrated in panel 2b of Fig. 8.1 by taking the Pauli blocking into account, while the electron and the hole interact via the screened *Coulomb* potential. In panel 2b of Fig. 8.2 we show how an increasing free-electron concentration in this case leads to a larger optical gap since the BMS [193] causes the absorption onset to occur at higher photon energies.

We observe another remarkable effect related to a Fermi-edge singularity (FES) [195] of the absorption: As the free-electron density increases, an excitonic bound state still occurs at the Fermi edge and its oscillator strength *increases* as well. This unphysical behavior contradicts experimental results [97]. However, it is not an artifact of the two-band model as we show later. Moreover, this occurs due to the sharpness of the Fermi surface at $T = 0$ K. The more free carriers occur in the system, the larger is the Fermi surface and the higher is the peak at the absorption onset since the impact of the degenerate electron gas on the electron-hole interaction, i.e., an additional screening, is missing.

Combining the modified electron-hole interaction and the Pauli blocking

Finally, as depicted in panel 3 of Fig. 8.1, we take both the Pauli blocking and the Yukawa potential for the electron-hole interaction into account. In panel 3 of Fig. 8.2 the resulting absorption coefficient is shown for the free-electron densities $n_{c,1}$ and $n_{c,2}$. Aside from the BMS, in both cases we observe a small peak at the absorption edge, which has been traced back to a logarithmic FES for this model [195], the so-called *Mahan exciton*. It is associated with a bound excitonic state with an excitation energy that is *lower* than the BMS-shifted QP gap. Due to the modified electron-hole interaction in the presence of a degenerate electron gas, we find that this singularity is much less pronounced than found before (cf. panel 2b of Fig. 8.2). In an analytical study of the two-band model [195] G. Mahan discovered in the 1960's the occurrence of these Mahan excitons as bound excitonic states for *all* finite free-electron concentrations. They emerge due to the effective electron-hole interaction potential which results when the Pauli blocking and the Yukawa potential are combined.

Influence of a temperature-dependent Fermi distribution of the free electrons

So far we studied merely the $T = 0$ K case and, therefore, assumed a step-function for the occupation numbers of the lowest CB. To account for finite temperatures we employ a temperature-dependent Fermi distribution for the occupation numbers of the CB states, i.e.,

$$n_{c\mathbf{k}} = \left[\exp\left(\frac{\varepsilon_{c\mathbf{k}}^{QP} - \varepsilon_F}{k_B T}\right) + 1\right]^{-1}. \tag{8.1}$$

In panel 3 of Fig. 8.2 we show the results for the absorption at $T = 300$ K. From this figure it becomes clear that the curves calculated for room temperature show an additional broadening of the absorption onset which significantly smoothens the FES.

8.2 *Ab-initio* calculations for *wz*-ZnO

In the preceding section we disentangled two different aspects of free electrons and their impact on the optical absorption for *wz*-ZnO. Within the two-band model only one VB and one CB contribute, while in the *ab-initio* approach introduced in Section 2.4 the Coulomb interaction couples electron-hole states of all bands in the excitonic Hamiltonian, Eq. (2.62). In addition, we assumed the matrix elements for the optical transitions within the two-band model to be independent of **k** [cf. Eq. (2.80)] in order to keep the calculations simple, which is an approximation that is only valid in the direct vicinity of Γ. For these two reasons it is *a priori* not clear to which extent the results of the two-band model are also quantitatively valid for a real semiconductor, hence, we aim for a full *ab-initio* treatment of the problem by accounting for the Pauli blocking as well as the modified electron-hole interaction when setting up the excitonic Hamiltonian, Eq. (2.62). We are then able to calculate the DF as well as the lowest eigenvalues and eigenstates of the excitonic Hamiltonian which we computed from a GGA+U+Δ starting electronic structure (cf. Section 3.1.2). The lowest CB states occur at Γ and its vicinity and we are able to resolve free-electron densities of about 10^{17} cm^{-3} using a fine sampling of this region of the BZ by means of hybrid **k**-point meshes (cf. Section 3.2). We incorporate the Pauli blocking in the BSE for the polarization function, Eq. (2.59), by introducing **k**-dependent occupation numbers $n_{c\mathbf{k}}$ for the lowest CB [57]. In addition, the modified electron-hole interaction is incorporated into the Hamiltonian, Eq. (2.62), by replacing the screened Coulomb potential with the Yukawa potential [cf. Eq. (2.73)]. The TF wave vector q_{TF}, Eq. (2.71), is calculated for each respective free-electron density n_c in the CB. The intraband contributions to the DF due to the free electrons are taken into account for the screening of the electron-hole interaction, i.e., for the *microscopic* DF (cf. Section 2.4.4).

8.2 Ab-initio calculations for wz-ZnO

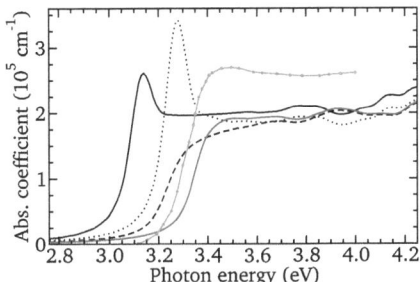

Figure 8.3: Absorption coefficient of wz-ZnO versus photon energy. We compare the curves for the undoped material (solid black) to results for doped ZnO with a free-electron concentration of $n_{c,1} = 1.9 \cdot 10^{19}$ cm^{-3}, calculated with only the Pauli blocking (dotted black), only the modified electron-hole interaction (dashed black) or both effects simultaneously (dark gray). An experimental result from Ref. [97] is also given (light gray).

An additional intraband contribution to the *macroscopic* DF, Eq. (2.67), is neglected since it affects only the infrared spectral range, whereas we focus on the optical absorption onset of wz-ZnO.

8.2.1 Absorption coefficient

We employ the complex DF $\varepsilon(\omega)$ calculated from the modified excitonic Hamiltonian to compute the absorption coefficient $\alpha(\omega)$ for wz-ZnO by means of the expression

$$\alpha(\omega) = \frac{\sqrt{2}\omega}{c} \cdot \sqrt{|\varepsilon(\omega)| - \text{Re}\,\varepsilon(\omega)}. \qquad (8.2)$$

In Fig. 8.3 the results for the undoped material are compared to the different approaches of incorporating the influence of the degenerate electron gas, i.e., when only the Yukawa electron-hole interaction (cf. panel 2a in Fig. 8.1), only the Pauli blocking (cf. panel 2b in Fig. 8.1), or both effects simultaneously (cf. panel 3 in Fig. 8.1) were taken into account. The results qualitatively agree with the trends predicted by the two-band model. As before, including merely the Pauli blocking leads to an unphysical increase of the oscillator strength, whereas the modified electron-hole interaction causes an extinction of the excitonic bound state at the absorption edge. Again, the combination of both effects results in a curve with a significantly modified line shape of the absorption onset due to the occurrence of a Mahan exciton.

In addition, in Fig. 8.3 we compare our results to a curve from a measurement of the absorption coefficient of doped wz-ZnO samples by T. Makino et al. [97]. We find perfect agreement of the line shape around the absorption onset of the measured curve for a sample with a free-electron concentration $n_{c,1}$ and our calculated curve that accounts for the Pauli blocking as well as the modified electron-hole interaction. We emphasize that the experimental result can only be explained by taking both effects into account. The higher absorption

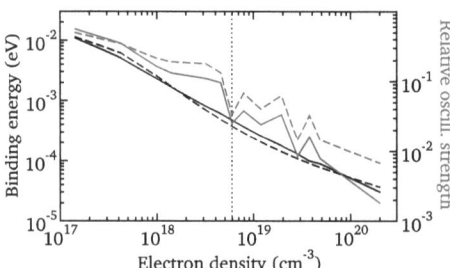

Figure 8.4: Exciton binding energies (solid curves) and oscillator strengths (dashed curves, normalized to the value for the undoped material) versus free-electron concentration as obtained from the solution of the BSE for wz-ZnO. Black curves result from calculations where the Yukawa potential has been used to model the electron-hole interaction (panel 2a in Fig. 8.1), whereas the Pauli blocking has been additionally taken into account for the gray curve (panel 3 in Fig. 8.1). The black dotted line represents the Mott density n_M (see text).

in experiment can be due to imperfect samples, whereas our calculations describe defect-free crystals.

Furthermore, we want to remark that in the preceding discussion the influence of the degenerate electron gas on the QP energies has been neglected. The presence of free electrons leads to an additional shrinkage of the QP gap (*gap renormalization*) that approximately can be calculated using the formula given by K.-F. Berggren and B. Sernelius [196, 197], i.e.,

$$\Delta E_g^{FE} = -\frac{e^2 k_F}{2\pi^2 \varepsilon_0 \varepsilon_{\text{eff}}} - \frac{e^2 q_{TF}}{8\pi \varepsilon_0 \varepsilon_{\text{eff}}} \left[1 - \frac{4}{\pi} \arctan\left(\frac{k_F}{q_{TF}}\right)\right]. \quad (8.3)$$

From this expression we find that for a free-electron concentration $n_{c,1}$ ($n_{c,2}$) the shrinkage ΔE_g^{FE} would amount to -0.213 eV (-0.261 eV). The BMS shows with 0.137 eV (0.236 eV) the same order of magnitude but the opposite sign. We want to point out that the energetic positions of the absorption onset of our calculated curve and the measured result (cf. Fig. 8.3) agree so well because, by accident, the gap underestimation discussed in Section 4.1.1 and the gap renormalization ΔE_g^{FE} due to the free-electron density $n_{c,1}$ compensate each other.

8.2.2 Binding energies and oscillator strengths

As discussed before, the Mott transition of the exciton occurs when the electron-hole interaction potential has no bound state anymore. Many studies exist for the Yukawa potential, exploring the corresponding values for q_{TF} or for the free-electron density (see Ref. [66] and references therein), leading to Eq. (2.75). Using the theoretical values of the electronic static dielectric constant ε_{eff} and the reduced electron-hole pair mass μ, the exciton radius in wz-

8.2 Ab-initio calculations for wz-ZnO

ZnO follows from Eq. (2.74), $a_{B,exc} = 1.24$ nm. With the CB mass m_c, Eq. (2.75) yields a Mott density of $n_M = 5.94 \cdot 10^{18}$ cm^{-3} for wz-ZnO (cf. Fig 8.4). Other formulae or "rules of thumb" were employed to calculate n_M, leading to values that are two orders of magnitude smaller or larger [192].

To analyze this situation by means of parameter-free calculations, we plot the largest exciton binding energies [difference of the lowest non-interacting electron-hole pair energy (including BMS) and the lowest eigenvalue of the excitonic Hamiltonian] and the corresponding optical oscillator strength as a function of the free-electron concentration in Fig. 8.4. We find a significant decrease in the binding energy as well as the oscillator strength with an increasing concentration of free electrons, whether or not the Pauli blocking is included in the calculations. Figure 8.4 also shows that taking the Pauli blocking into account leads to an increase in the binding energies and the oscillator strengths with respect to the situation where we only account for the modified electron-hole interaction. This effect is especially large for free-electron densities between 10^{18} cm^{-3} and 10^{19} cm^{-3}. Moreover, we point out that already for densities of about 10^{19} cm^{-3} the exciton binding energy is below 1 meV and, therefore, very hard to be observed in *ab-initio* calculations (due to the high computational cost for achieving the necessary convergence) as well as in experiments. Since the Pauli blocking is taken into account by modifying the **k**-dependent occupation numbers, the corresponding binding energies are very sensitive to the **k** sampling of the BZ. As a consequence, we observe numerical fluctuations of the binding energies in the sub-meV range for $n > 3 \cdot 10^{18}$ cm^{-3} (cf. Fig. 8.4), noting that such small values constitute a challenge for the theoretical description. In contrast, the curve calculated without Pauli blocking does not suffer from this sampling problem. However, it does not show any indication of a Mott transition at n_M, Eq. (2.75), which demonstrates that our numerical approach cannot resolve such small binding energies correctly.

According to G. Mahan [195] the Mott transition of the exciton does not occur in a system with a filled VB and additional free electrons in the lowest CB since the Pauli blocking modifies the electron-hole interaction in such a way that the bound state does not vanish for finite free-electron concentrations. However, the oscillator strengths as well as the exciton binding energies decrease very rapidly with increasing free-electron concentrations (cf. Fig. 8.4). At $n_c = 4.5 \cdot 10^{18}$ cm^{-3} the oscillator strength amounts to only about 13 % of the value for the undoped material. Above $n_c \approx 5 \cdot 10^{19}$ cm^{-3} it falls below 1 % of the original value. This continuous decrease instead of a clear criterion explains the high uncertainty of experimentally observed Mott densities.

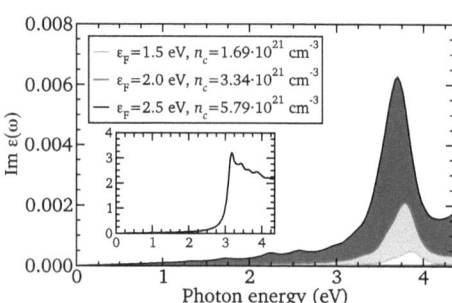

Figure 8.5: Contributions of the inter-conduction-band absorption to the imaginary part of the dielectric function for three different free-electron concentrations in the lowest conduction band of wz-ZnO. These curves were calculated within independent-particle approximation and averaged over parallel and perpendicular polarization. For all three cases the lowest valence-conduction band excitation occurs at higher energies than shown in the plot. For comparison the inset shows the BSE curve discussed in Section 4.2.2.

8.2.3 Inter-conduction-band absorption

In addition to the effects that have been discussed so far, the electrons that occupy the lowest CB states can also absorb photons and be optically excited into higher lying CBs. This, in turn, might limit band-gap tuning via doping [98] since the additional transitions from the lowest occupied CB into higher empty ones can occur below the lowest VB-CB gap in a doped system. Therefore, we calculate the DF of non-interacting electron-hole pairs from the GGA+U+Δ electronic structure to arrive at an estimate for the order of magnitude of this effect.

In Fig. 8.5 we plot the contributions to the DF that arise due to transitions from the occupied lowest CB states into higher CBs. This figure indicates that the overall effect is small, even for relatively high free-electron concentrations. We attribute this to the small optical transition-matrix elements that we find for the corresponding transitions, since they are dipole-forbidden for symmetry reasons at the Γ point of the BZ for wz-ZnO. In the vicinity of the Γ point, these matrix elements tend to increase and, therefore, the peak heights of the curves shown in Fig. 8.5 increase with the free-electron density. Even though their absolute values remain small, these contributions to the absorption spectrum still need to be kept in mind since they can be important, especially for highly precise measurements of the absorption coefficient of doped ZnO.

8.3 Summary

In this chapter we studied the Pauli blocking and the modification of the electron-hole interaction due to additional screening; both occur when a degenerate electron gas is present in the lowest CB of *wz*-ZnO. We found that their combination allows for an accurate description of the line shape of a measured absorption spectrum of *n*-doped ZnO by means of *ab-initio* calculations. In addition, we observed that a temperature-dependent Fermi-occupation of the lowest CB states introduces an additional broadening of the absorption edge with respect to the result for $T = 0$ K.

From a calculation of the exciton binding energy of the lowest bound electron-hole pair state, along with the corresponding oscillator strength, we explained why a Mott transition of the exciton can barely be observed in experiment. The continuous decrease of both quantities with an increasing free-electron density renders it difficult to determine the Mott density.

Furthermore, the additional free electrons cause an inter-conduction-band contribution to the absorption. While the respective peaks in the DF can occur below the smallest VB-CB gap, the absolute magnitude of this effect remains small.

9 The end ... and future prospects

> In the end we shall have had enough of cynicism and skepticism and humbug and we shall want to live more musically.
>
> Vincent van Gogh

Most advanced theoretical-spectroscopy techniques that deal with one- and two-particle excitations in semiconductors have been employed and extended to unravel the physics of a variety of imperfections of the group-II oxides MgO, ZnO, and CdO. We explored how the impact of real-structure effects contributes to the possibility of combining ZnO with MgO and CdO in the context of materials design — paving the way towards transparent oxide electronics. For that purpose we applied *electronic-structure* calculations based on the HSE hybrid functional, as sophisticated approximation to exchange and correlation, which provides a well-suited starting point for the computation of quasiparticle energies within Hedin's *GW* approximation. For a parameter-free description of *optical properties*, we incorporated the electron-hole interaction into the calculations, via the solution of a Bethe-Salpeter equation, to account for excitonic and local-field effects.

We proved the high accuracy and reliability of these approaches by comparing quasiparticle band structures, densities of states, spin-orbit splittings, effective band masses as well as the complex dielectric functions and the electron-energy loss functions for the ideal equilibrium polymorphs *rs*-MgO, *wz*-ZnO, and *rs*-CdO to experimental results. The reassuring agreement found for all three oxides provides confidence in our methods and justifies their suitability to be applied to real-structure effects.

The ordering of the uppermost valence bands of *wz*-ZnO has been debated for decades. Using parameter-free calculations we were able to thoroughly answer this question even in the presence of uniaxial and biaxial strain that exceeds experimentally accessible magnitudes.

We studied pseudobinary $Mg_xZn_{1-x}O$ and $Cd_xZn_{1-x}O$ alloys by combining the different crystal structures of the binary end components within one cluster expansion. From a comparison of the thermodynamic equilibrium situation for the isostructural as well as the heterostructural systems to growth conditions where preferably the binary end components occur or where the clusters are purely stochastically distributed we understood trends of different experimental findings for composition ranges in which either the *wz* or the *rs* crystal structure dominates. We achieved the computationally extremely demanding calculation of the quasiparticle energies for *all cluster cells*. We derived the bowing of the fundamental gaps, which reveals a strongly nonlinear behavior, and determined configurational averages for densities of states. Calculating these averages for the dielectric functions of wz-$Mg_xZn_{1-x}O$ alloys based on results that include excitonic effects for *all clusters* of the expansion has never been achieved before.

Furthermore, we have demonstrated for two charge states of the important F-center of rs-MgO that the strong resemblance of the defect-related optical absorption peaks, that has been puzzling for decades, can be understood by combining sophisticated electronic-structure calculations with the computation of exciton binding energies. Both the solution of the Bethe-Salpeter equation for the 63-atom defect supercells and the inclusion of spin-polarization render these calculations highly expensive. Hence, they are not routinely performed today.

For the first time the *ab-initio* description of electron-hole excitations has been applied to study free electrons as they are present for instance in the lowest conduction band of heavily *n*-doped wz-ZnO. This approach allows for the accurate description of the Mahan exciton which forms at the optical absorption edge. The temperature dependence of the Fermi function of the lowest conduction-band states leads to an additional broadening which destroys the Fermi-edge singularity. We explained why the Mott transition of the exciton can hardly be observed in experiment and theory. In addition, inter-conduction-band absorption due to the free electrons can occur below the Burstein-Moss shifted absorption edge and limit band-gap tailoring by means of doping.

In this work we employed sophisticated approaches based on many-body perturbation theory to successfully describe electronic excitations in imperfect crystals of transparent conductive oxides. Aside from the success of the methods that we applied to tackle systems of large complexity, our results suggest that the consistent inclusion of the electron-phonon coupling in the parameter-free approach will have to be one of the next steps. In addition, this work undoubtedly proves the need for taking real-structure effects into account in *ab-initio* calculations. Henceforth, such developments will experience a boost and will more and more enable mankind to benefit from computer-aided *materials* design.

A Appendix

A.1 Cluster expansions for the wurtzite and the rocksalt crystal structure

In Fig. A.1 we depict the 16-atom cluster cells that we use for the cluster expansions of the alloys in the *wz* and in the *rs* crystal structure. All possible classes are given in Table A.1 along with the respective numbers of cations and their positions in the cell (cf. Fig. A.1). While the expansion for the *wz* crystal structure was published in Ref. [74], the expansion for the *rs* crystal structure was developed in collaboration with the diploma student Matthias Eisenacher [76, 198], who was supervised by the author of this work.

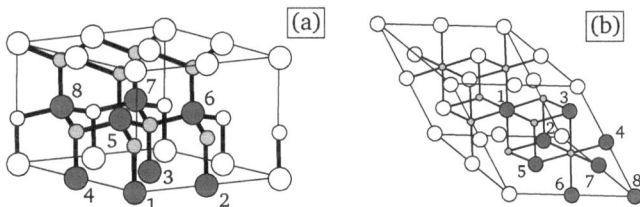

Figure A.1: Illustration of atomic sites in the 16-atom clusters of the wurtzite (a) and the rocksalt (b) structure. Anions (C atoms) are depicted in light gray (small), cations (A or B atoms) in dark gray (large). White atoms represent the surrounding lattice and are not part of the 16-atom clusters. The denotation 1...8 of the cations is used in Table A.1.

A.2 Parameters used in the calculations

For studying the group-II oxides we treat the Mg $2p$ and Mg $3s$ as well as the O $2s$ and O $2p$ electrons as valence electrons. For Zn and Cd, the Zn $3d$ and Cd $4d$ states have to be included as valence states, respectively, together with the Zn $4s$ and Cd $5s$ electrons. For the calculations related to the oxygen vacancy in MgO we take only the Mg $3s$ electrons into account, since these calculations are based on input from Patrick Rinke.

	wz				rs		
j	n_j	g_j	A atoms	j	n_j	g_j	A atoms
0	0	1	–	22	0	1	–
1	1	8	1	23	1	8	1
2	2	12	1,2	24	2	24	1,2
3	2	12	1,5	25	2	4	4,5
4	2	4	1,8	26	3	32	1,2,3
5	3	8	1,2,3	27	3	24	1,4,5
6	3	24	1,2,7	28	4	8	1,2,3,4
7	3	24	1,2,5	29	4	8	1,2,3,5
8	4	2	1,2,3,4	30	4	48	1,2,4,5
9	4	8	1,2,3,5	31	4	6	3,4,5,6
10	4	24	1,2,4,5	32	5	32	1,2,3,4,5
11	4	6	1,2,5,6	33	5	24	1,3,4,5,6
12	4	6	1,2,7,8	34	6	24	1,2,3,4,5,6
13	4	24	1,2,5,8	35	6	4	2,3,4,5,6,7
14	5	24	3,4,6,7,8	36	7	8	1,2,3,4,5,6,7
15	5	24	3,4,5,6,8	37	8	1	1,2,3,4,5,6,7,8
16	5	8	4,5,6,7,8				
17	6	4	2,3,4,5,6,7				
18	6	12	2,3,4,6,7,8				
19	6	12	3,4,5,6,7,8				
20	7	8	2,3,4,5,6,7,8				
21	8	1	1,2,3,4,5,6,7,8				

Table A.1: Cluster classes for 16-atom cells in wz- and rs-crystal structure. For each class j the number n_j of A cations, the degeneracy g_j of the class, and the cation sites occupied with A-type atoms (for one representative of the class) are given.

Exchange-correlation functional

Throughout this work we employ the PW GGA II functional by Y. Wang and J. Perdew [16, 21], also called PW91. Only for the calculations related to the oxygen vacancy in MgO (cf. Chapter 7) the LDA [15] is used to be consistent with the input from Patrick Rinke.

Cutoff energy of the plane-wave expansion

For the equilibrium polymorphs of the group-II oxides, i.e., rs-MgO, wz-ZnO, and rs-CdO we found earlier [19, 23] that an energy cutoff of 400 eV is suitable to yield converged results. The same value has been proven to be well-suited to study uniaxially and biaxially strained wz-ZnO, as well as the non-equilibrium compounds wz-MgO and wz-CdO. In the case of the

A.2 Parameters used in the calculations

	type of calculation	k mesh
rs-MgO (2 atom cell)	HSE+G_0W_0	Γ-centered, $16 \times 16 \times 16$
rs-MgO (2 atom cell)	HSE+SOC	Γ-centered, $10 \times 10 \times 10$
wz-ZnO (4 atom cell)	HSE+G_0W_0	Γ-centered, $8 \times 8 \times 6$
wz-ZnO (4 atom cell)	HSE+SOC	Γ-centered, $8 \times 8 \times 6$
rs-CdO (2 atom cell)	HSE+G_0W_0	Γ-centered, $16 \times 16 \times 16$
rs-CdO (2 atom cell)	HSE+SOC	Γ-centered, $10 \times 10 \times 10$
strained wz-ZnO (4 atom cell)	HSE+G_0W_0	Γ-centered, $8 \times 8 \times 6$
strained wz-ZnO (4 atom cell)	HSE+SOC	Γ-centered, $8 \times 8 \times 6$
alloys, wz crystal structure (16 atom cells)	ground-state	$2 \times 2 \times 2$
alloys, rs crystal structure (16 atom cells)	ground-state	$4 \times 4 \times 4$
alloys (16 atom cells)	HSE+G_0W_0	Γ-centered, $3 \times 3 \times 3$
alloys (16 atom cells)	HSE+SOC	Γ-centered, $3 \times 3 \times 3$

Table A.2: The MP **k**-point meshes that have been used to sample the Brillouin zones in the different calculations.

pseudobinary alloys we found that the cutoff energy has to be increased to 450 eV in all cases to converge our results.

Sampling of the Brillouin zone

For the calculation of the QP energies and band structures as well as the HSE band structures including the spin-orbit interaction we employ the **k**-point meshes given in Table A.2.

The computation of the DFs for the equilibrium polymorphs of the group-II oxides, i.e., rs-MgO, wz-ZnO, and rs-CdO, is divided into two different calculations: one for the low-energy region and one for higher photon energies (cf. Section 3.2). For rs-MgO, photon energies up to $\hbar\omega = 13.4$ eV are included via a $10:5:30$ hybrid mesh and for energies up to $\hbar\omega = 32.5$ eV a regular $10 \times 10 \times 10$ MP mesh is employed. In the case of wz-ZnO, transitions up to 8.5 eV are computed using a hybrid $8 \times 8 \times 6 : 5 \times 5 \times 5 : 20.8 \times 20.8 \times 15.6$ mesh and the high-energy part is calculated using regular $6 \times 6 \times 4$ MP points. Because CdO is an indirect semiconductor, it is not sufficient to sample only a certain region in **k** space, e.g. around the Γ point, at higher density. Instead, we employ two regular MP meshes: $18 \times 18 \times 18$ **k** points for transitions up to 7.8 eV and $8 \times 8 \times 8$ points up to an energy of 32.5 eV.

In the case of the pseudobinary wz-Mg$_x$Zn$_{1-x}$O alloys we calculate the DF up to a photon energy of 5.8 eV (*before* applying the scissors shifts given in Table A.3) using $9 \times 9 \times 9$ **k** points. For higher photon energies we employ a $4 \times 4 \times 4$ mesh. In both cases the meshes are shifted by a small random vector, as discussed earlier (cf. Section 3.2).

To study the influence of free electrons on the optical properties of wz-ZnO we significantly

j	Δ_j	j	Δ_j	j	Δ_j	j	Δ_j
0	1.46	6	1.92	12	2.16	18	2.39
1	1.60	7	1.83	13	2.02	19	2.29
2	1.70	8	1.91	14	2.11	20	2.62
3	1.72	9	1.96	15	2.25	21	2.78
4	1.79	10	1.99	16	2.11	–	–
5	1.82	11	1.92	17	2.48	–	–

Table A.3: Scissors operators used for the individual clusters of the cluster expansion for the wz-$Mg_xZn_{1-x}O$ alloys. The values are obtained using the HSE+G_0W_0 and the GGA+U band structures of each individual cluster.

increase the sampling in the vicinity of the Γ point by employing a $8 \times 8 \times 6 : 3 \times 3 \times 3 : 24 \times 24 \times 18 : 5 \times 5 \times 5 : 96 \times 96 \times 72$ double-hybrid **k**-point mesh [61].

Lattice parameters

In order to obtain the equilibrium values of the cell volumes, total energies, and bulk moduli we minimize the total energy for several cell volumes and fit the resulting volume dependence of the total energy to the Murnaghan equation of state [199]. In each case, we compute fully relaxed atomic positions, i.e., optimized lattice constants and internal cell parameters, ensuring that the forces acting on the ions are well below 5 meV/Å.

The lattice parameters we employ in this work are summarized in Table A.4. The atomic geometries for the defect supercell containing the F-center in rs-MgO (cf. Table A.4) were also provided by Patrick Rinke and have been determined within LDA. We do not give the cell parameters for each of the 76 cluster cells for the studies of the pseudobinary alloys here, but we refer to the information about the cation-cation distances as compiled in Ref. [76].

Dielectric constants

The model function that is employed for the screening of the electron-hole interaction in the calculation of the excitonic Hamiltonian needs the effective dielectric constant as input parameter. Within this work we employ $\varepsilon_{\text{eff}} = 3.0$ for rs-MgO, which is close to both the experimental value of 2.94 and the result of 3.16 that we obtained within independent-particle approximation. For wz-ZnO we employ the result obtained within the GGA+U approximation, averaged over both polarization directions, i.e., $\varepsilon_{\text{eff}} = 4.4$. In the case of rs-CdO the GGA+U result amounts to $\varepsilon_{\text{eff}} = 6.55$.

Also for the pseudobinary wz-$Mg_xZn_{1-x}O$ alloys we determined the screening constant within the GGA+U approach for each cluster individually. These values are compiled in Table A.5.

A.2 Parameters used in the calculations

	a_0 (Å)	a (Å)	c (Å)	u
rs-MgO	4.254	–	–	–
wz-ZnO	–	3.283	5.309	0.379
rs-CdO	4.779	–	–	–
wz-ZnO, $\varepsilon_u = -0.02$	–	3.304	5.203	0.385
wz-ZnO, $\varepsilon_u = -0.01$	–	3.294	5.256	0.382
wz-ZnO, $\varepsilon_u = 0.01$	–	3.274	5.362	0.376
wz-ZnO, $\varepsilon_u = 0.02$	–	3.264	5.415	0.373
wz-ZnO, $\varepsilon_b = -0.02$	–	3.218	5.403	0.369
wz-ZnO, $\varepsilon_b = -0.01$	–	3.250	5.356	0.374
wz-ZnO, $\varepsilon_b = 0.01$	–	3.316	5.256	0.383
wz-ZnO, $\varepsilon_b = 0.02$	–	3.349	5.206	0.388
wz-MgO	–	3.322	5.136	0.392
wz-CdO	–	3.678	5.825	0.385
rs-MgO, (F-center, F$^+$-center)	4.151	–	–	–

Table A.4: Lattice parameters a_0 (for the rs crystal structure) as well as a and c (for the wz crystal structure). For wz crystals also the dimensionless internal cell parameter u is given.

j	$\varepsilon_{\text{eff},j}$	j	$\varepsilon_{\text{eff},j}$	j	$\varepsilon_{\text{eff},j}$	j	$\varepsilon_{\text{eff},j}$
0	4.25	6	3.67	12	3.51	18	3.25
1	4.03	7	3.67	13	3.52	19	3.25
2	3.85	8	3.54	14	3.38	20	3.12
3	3.84	9	3.53	15	3.38	21	3.00
4	3.83	10	3.52	16	3.39	–	–
5	3.69	11	3.52	17	3.25	–	–

Table A.5: Electronic static dielectric constants of the individual clusters of the cluster expansion for the wz-Mg$_x$Zn$_{1-x}$O alloy. The values are calculated within the GGA+U approach in the independent-particle approximation (IPA).

Modified *GW* implementation for the alloys

Before the calculations for the alloys were performed, Prof. Georg Kresse suggested a modified version of the *GW* implementation within the PAW scheme. By restoring the higher moments of the all-electron charge density also for its description on the plane-wave grid its shape is very accurately reproduced. While this does not influence the HSE results, it has an impact on the G_0W_0 calculations, since they lack (for computational reasons) an exact implementation of the one-center terms of the PAW approach [47, 48]. With the new approach the fundamental band gap of wz-ZnO is about 0.29 eV smaller than discussed in Chapter 4. For MgO and CdO as well as the rs polymorph of ZnO the influence has been found to be not larger than 0.1 eV.

Bibliography

[1] Z. L. Wang and J. Song: Piezoelectric Nanogenerators Based on Zinc Oxide Nanowire Arrays. In: *Science* **312** (5771), pp. 242–246 (2006), DOI: `10.1126/science.1124005`.

[2] M. Law, L. E. Greene, J. C. Johnson, R. Saykally, and P. Yang: Nanowire dye-sensitized solar cells. In: *Nature Materials* **4** (6), pp. 455–459 (2005), DOI: `10.1038/nmat1387`.

[3] P. Hohenberg and W. Kohn: Inhomogeneous Electron Gas. In: *Phys. Rev.* **136** (3B), B864–B871 (Nov. 1964), DOI: `10.1103/PhysRev.136.B864`.

[4] W. Kohn and L. J. Sham: Self-Consistent Equations Including Exchange and Correlation Effects. In: *Phys. Rev.* **140** (4A), A1133–A1138 (Nov. 1965), DOI: `10.1103/PhysRev.140.A1133`.

[5] L. Hedin: New Method for Calculating the One-Particle Green's Function with Application to the Electron-Gas Problem. In: *Phys. Rev.* **139** (3A), A796–A823 (Aug. 1965), DOI: `10.1103/PhysRev.139.A796`.

[6] L. Hedin and S. Lundqvist: „Effects of Electron-Electron and Electron-Phonon Interactions on the One-Electron States of Solids", in: *Advances in Research and Applications*, ed. by D. T. Frederick Seiz and H. Ehrenreich, vol. 23, Solid State Physics, Academic Press, 1970, pp. 1–181, DOI: `10.1016/S0081-1947(08)60615-3`.

[7] M. Born and J. R. Oppenheimer: Zur Quantentheorie der Molekeln. In: *Ann. Phys.* **389** (20), pp. 457–484 (1927), DOI: `10.1002/andp.19273892002`.

[8] P. A. M. Dirac: The Quantum Theory of the Electron. In: *P. Roy. Soc. Lond. A Mat.* **117** (778), pp. 610–624 (1928), URL: `http://www.jstor.org/stable/94981`.

[9] P. A. M. Dirac: The Quantum Theory of the Electron. Part II. In: *P. Roy. Soc. Lond. A Mat.* **118** (779), pp. 351–361 (1928), URL: `http://www.jstor.org/stable/94909`.

[10] W. Pauli jr.: Zur Quantenmechanik des magnetischen Elektrons. In: *Z. Phys. A* **43** (9), p. 601 (1927), DOI: `10.1007/BF01397326`.

[11] W. Nolting: *Grundkurs Theoretische Physik – Vielteilchentheorie*. Vol. 7, Springer, 2005.

[12] W. Kohn: Nobel Lecture: Electronic structure of matter—wave functions and density functionals. In: *Rev. Mod. Phys.* **71** (5), pp. 1253–1266 (Oct. 1999), DOI: 10.1103/RevModPhys.71.1253.

[13] R. M. Dreizler and E. K. Gross: *Density functional theory: an approach to the quantum many-body problem*. Springer, 1990.

[14] F. Bloch: Über die Quantenmechanik der Elektronen in Kristallgittern. In: *Z. Phys. A* **52** (7–8), p. 555 (1929), DOI: 10.1007/BF01339455.

[15] D. M. Ceperley and B. J. Alder: Ground State of the Electron Gas by a Stochastic Method. In: *Phys. Rev. Lett.* **45** (7), pp. 566–569 (Aug. 1980), DOI: 10.1103/PhysRevLett.45.566.

[16] J. P. Perdew, J. A. Chevary, S. H. Vosko, K. A. Jackson, M. R. Pederson, D. J. Singh, and C. Fiolhais: Atoms, molecules, solids, and surfaces: Applications of the generalized gradient approximation for exchange and correlation. In: *Phys. Rev. B* **46** (11), pp. 6671–6687 (Sept. 1992), DOI: 10.1103/PhysRevB.46.6671.

[17] P. Rinke, A. Janotti, M. Scheffler, and C. G. Van de Walle: Defect Formation Energies without the Band-Gap Problem: Combining Density-Functional Theory and the *GW* Approach for the Silicon Self-Interstitial. In: *Phys. Rev. Lett.* **102** (2), p. 026402 (Jan. 2009), DOI: 10.1103/PhysRevLett.102.026402.

[18] R. W. Godby, M. Schlüter, and L. J. Sham: Accurate Exchange-Correlation Potential for Silicon and Its Discontinuity on Addition of an Electron. In: *Phys. Rev. Lett.* **56** (22), pp. 2415–2418 (June 1986), DOI: 10.1103/PhysRevLett.56.2415.

[19] A. Schleife, F. Fuchs, J. Furthmüller, and F. Bechstedt: First-principles study of ground- and excited-state properties of MgO, ZnO, and CdO polymorphs. In: *Phys. Rev. B* **73** (24), p. 245212 (June 2006), DOI: 10.1103/PhysRevB.73.245212.

[20] J. P. Perdew and A. Zunger: Self-interaction correction to density-functional approximations for many-electron systems. In: *Phys. Rev. B* **23** (10), pp. 5048–5079 (May 1981), DOI: 10.1103/PhysRevB.23.5048.

[21] Y. Wang and J. P. Perdew: Correlation hole of the spin-polarized electron gas, with exact small-wave-vector and high-density scaling. In: *Phys. Rev. B* **44** (24), pp. 13298–13307 (Dec. 1991), DOI: 10.1103/PhysRevB.44.13298.

[22] J. P. Perdew and Y. Wang: Accurate and simple analytic representation of the electron-gas correlation energy. In: *Phys. Rev. B* **45** (23), pp. 13244–13249 (June 1992), DOI: 10.1103/PhysRevB.45.13244.

[23] A. Schleife: Ab-initio Untersuchungen elektronischer und optischer Eigenschaften von Gruppe-II-Oxiden. Diploma thesis, Jena: Friedrich-Schiller-Universität, 2006.

[24] V. I. Anisimov, J. Zaanen, and O. K. Andersen: Band theory and Mott insulators: Hubbard U instead of Stoner I. In: *Phys. Rev. B* **44** (3), pp. 943–954 (July 1991), DOI: 10.1103/PhysRevB.44.943.

[25] S. L. Dudarev, G. A. Botton, S. Y. Savrasov, C. J. Humphreys, and A. P. Sutton: Electron-energy-loss spectra and the structural stability of nickel oxide: An LSDA+U study. In: *Phys. Rev. B* **57** (3), pp. 1505–1509 (Jan. 1998), DOI: 10.1103/PhysRevB.57.1505.

[26] J. Hubbard: Electron Correlations in Narrow Energy Bands. In: *P. Roy. Soc. Lond. A Mat.* **276** (1365), pp. 238–257 (1963), URL: http://www.jstor.org/stable/2414761.

[27] V. I. Anisimov, I. V. Solovyev, M. A. Korotin, M. T. Czyżyk, and G. A. Sawatzky: Density-functional theory and NiO photoemission spectra. In: *Phys. Rev. B* **48** (23), pp. 16929–16934 (Dec. 1993), DOI: 10.1103/PhysRevB.48.16929.

[28] A. I. Liechtenstein, V. I. Anisimov, and J. Zaanen: Density-functional theory and strong interactions: Orbital ordering in Mott-Hubbard insulators. In: *Phys. Rev. B* **52** (8), R5467–R5470 (Aug. 1995), DOI: 10.1103/PhysRevB.52.R5467.

[29] A. Seidl, A. Görling, P. Vogl, J. A. Majewski, and M. Levy: Generalized Kohn-Sham schemes and the band-gap problem. In: *Phys. Rev. B* **53** (7), pp. 3764–3774 (Feb. 1996), DOI: 10.1103/PhysRevB.53.3764.

[30] C. Adamo and V. Barone: Toward reliable density functional methods without adjustable parameters: The PBE0 model. In: *J. Chem. Phys.* **110** (13), pp. 6158–6170 (1999), DOI: 10.1063/1.478522.

[31] M. Ernzerhof and G. E. Scuseria: Assessment of the Perdew–Burke–Ernzerhof exchange-correlation functional. In: *J. Chem. Phys.* **110** (11), pp. 5029–5036 (1999), DOI: 10.1063/1.478401.

[32] J. P. Perdew, K. Burke, and M. Ernzerhof: Generalized Gradient Approximation Made Simple. In: *Phys. Rev. Lett.* **77** (18), pp. 3865–3868 (Oct. 1996), DOI: 10.1103/PhysRevLett.77.3865.

[33] A. D. Becke: A new mixing of Hartree–Fock and local density-functional theories. In: *J. Chem. Phys.* **98** (2), pp. 1372–1377 (1993), DOI: 10.1063/1.464304.

[34] J. P. Perdew, M. Ernzerhof, and K. Burke: Rationale for mixing exact exchange with density functional approximations. In: *J. Chem. Phys.* **105** (22), pp. 9982–9985 (1996), DOI: 10.1063/1.472933.

[35] J. Heyd, G. E. Scuseria, and M. Ernzerhof: Hybrid functionals based on a screened Coulomb potential. In: *J. Chem. Phys.* **118** (18), pp. 8207–8215 (2003), DOI: 10.1063/1.1564060.

[36] J. Heyd, G. E. Scuseria, and M. Ernzerhof: Erratum: "Hybrid functionals based on a screened Coulomb potential" [J. Chem. Phys. **118**, 8207 (2003)]. In: *J. Chem. Phys.* **124** (21), p. 219906 (2006), DOI: 10.1063/1.2204597.

[37] A. V. Krukau, O. A. Vydrov, A. F. Izmaylov, and G. E. Scuseria: Influence of the exchange screening parameter on the performance of screened hybrid functionals. In: *J. Chem. Phys.* **125** (22), p. 224106 (2006), DOI: 10.1063/1.2404663.

[38] J. Paier, M. Marsman, K. Hummer, G. Kresse, I. C. Gerber, and J. G. Ángyán: Screened hybrid density functionals applied to solids. In: *J. Chem. Phys.* **124** (15), p. 154709 (2006), DOI: 10.1063/1.2187006.

[39] J. Paier, M. Marsman, K. Hummer, G. Kresse, I. C. Gerber, and J. G. Ángyán: Erratum: "Screened hybrid density functionals applied to solids" [J. Chem. Phys. **124**, 154709 (2006)]. In: *J. Chem. Phys.* **125** (24), p. 249901 (2006), DOI: 10.1063/1.2403866.

[40] U. von Barth and L. Hedin: A local exchange-correlation potential for the spin polarized case. I. In: *J. Phys. C Solid State* **5** (13), pp. 1629–1642 (1972), DOI: 10.1088/0022-3719/5/13/012.

[41] A. H. MacDonald and S. H. Vosko: A relativistic density functional formalism. In: *J. Phys. C Solid State* **12** (15), pp. 2977–2990 (1979), DOI: 10.1088/0022-3719/12/15/007.

[42] D. Hobbs, G. Kresse, and J. Hafner: Fully unconstrained noncollinear magnetism within the projector augmented-wave method. In: *Phys. Rev. B* **62** (17), pp. 11556–11570 (Nov. 2000), DOI: 10.1103/PhysRevB.62.11556.

[43] M. S. Hybertsen and S. G. Louie: Electron correlation in semiconductors and insulators: Band gaps and quasiparticle energies. In: *Phys. Rev. B* **34** (8), pp. 5390–5413 (Oct. 1986), DOI: 10.1103/PhysRevB.34.5390.

[44] H. Stolz: *Einführung in die Vielelektronentheorie der Kristalle*. Akademie-Verlag, Berlin, 1974.

[45] S. G. Louie: *Predicting Materials and Properties: Theory of the Ground and Excited State*. Vol. Conceptual Foundations of Materials, Contemporary Concepts of Condensed Matter Science, Elsevier, 2006, chap. 2, pp. 9–53.

[46] W. G. Aulbur, L. Jönsson, and J. W. Wilkins: „Quasiparticle Calculations in Solids", in: *Advances in Research and Applications*, ed. by H. Ehrenreich and F. Spaepen, vol. 54, Solid State Physics, Academic Press, 1999, pp. 1–218, DOI: 10.1016/S0081-1947(08)60248-9.

[47] F. Fuchs, J. Furthmüller, F. Bechstedt, M. Shishkin, and G. Kresse: Quasiparticle band structure based on a generalized Kohn-Sham scheme. In: *Phys. Rev. B* **76** (11), p. 115109 (Sept. 2007), DOI: 10.1103/PhysRevB.76.115109.

[48] M. Shishkin and G. Kresse: Implementation and performance of the frequency-dependent *GW* method within the PAW framework. In: *Phys. Rev. B* **74** (3), p. 035101 (July 2006), DOI: 10.1103/PhysRevB.74.035101.

[49] P. Rinke, A. Qteish, J. Neugebauer, C. Freysoldt, and M. Scheffler: Combining *GW* calculations with exact-exchange density-functional theory: an analysis of valence-band photoemission for compound semiconductors. In: *New J. Phys.* **7**, p. 126 (2005), DOI: 10.1088/1367-2630/7/1/126.

[50] F. Fuchs: Ab-initio-Methoden zur Berechnung der elektronischen Anregungseigenschaften von Halbleitern und Isolatoren unter Berücksichtigung von Vielteilchen-Effekten. PhD thesis, Jena: Friedrich-Schiller-Universität, 2008, URL: http://www.db-thueringen.de/servlets/DocumentServlet?id=12532.

[51] G. Strinati: Application of the Green's functions method to the study of the optical properties of semiconductors. In: *La Rivista del Nuovo Cimento* **11** (12), pp. 1–86 (1988), DOI: 10.1007/BF02725962.

[52] P. Hahn: Berechnung von Vielteilcheneffekten in den Anregungsspektren von Kristallen, Oberflächen und Molekülen. PhD thesis, Jena: Friedrich-Schiller-Universität, 2004, URL: http://d-nb.info/973939079.

[53] S. L. Adler: Quantum Theory of the Dielectric Constant in Real Solids. In: *Phys. Rev.* **126** (2), pp. 413–420 (Apr. 1962), DOI: 10.1103/PhysRev.126.413.

[54] N. Wiser: Dielectric Constant with Local Field Effects Included. In: *Phys. Rev.* **129** (1), pp. 62–69 (Jan. 1963), DOI: 10.1103/PhysRev.129.62.

[55] G. Onida, L. Reining, and A. Rubio: Electronic excitations: density-functional versus many-body Green's-function approaches. In: *Rev. Mod. Phys.* **74** (2), pp. 601–659 (June 2002), DOI: 10.1103/RevModPhys.74.601.

[56] W. Hanke and L. J. Sham: Many-Particle Effects in the Optical Excitations of a Semiconductor. In: *Phys. Rev. Lett.* **43** (5), pp. 387–390 (July 1979), DOI: 10.1103/PhysRevLett.43.387.

[57] C. Rödl: Elektronische und exzitonische Anregungen in magnetischen Isolatoren. PhD thesis, Jena: Friedrich-Schiller-Universität, 2009.

[58] S. Albrecht, L. Reining, R. Del Sole, and G. Onida: Ab Initio Calculation of Excitonic Effects in the Optical Spectra of Semiconductors. In: *Phys. Rev. Lett.* **80** (20), pp. 4510–4513 (May 1998), DOI: 10.1103/PhysRevLett.80.4510.

[59] L. X. Benedict, E. L. Shirley, and R. B. Bohn: Optical Absorption of Insulators and the Electron-Hole Interaction: An Ab Initio Calculation. In: *Phys. Rev. Lett.* **80** (20), pp. 4514–4517 (May 1998), DOI: 10.1103/PhysRevLett.80.4514.

[60] M. Rohlfing and S. G. Louie: Electron-Hole Excitations in Semiconductors and Insulators. In: *Phys. Rev. Lett.* **81** (11), pp. 2312–2315 (Sept. 1998), DOI: 10.1103/PhysRevLett.81.2312.

[61] F. Fuchs, C. Rödl, A. Schleife, and F. Bechstedt: Efficient $\mathcal{O}(N^2)$ approach to solve the Bethe-Salpeter equation for excitonic bound states. In: *Phys. Rev. B* **78** (8), p. 085103 (Aug. 2008), DOI: 10.1103/PhysRevB.78.085103.

[62] P. H. Hahn, W. G. Schmidt, and F. Bechstedt: Bulk Excitonic Effects in Surface Optical Spectra. In: *Phys. Rev. Lett.* **88** (1), p. 016402 (Dec. 2001), DOI: 10.1103/PhysRevLett.88.016402.

[63] W. G. Schmidt, S. Glutsch, P. H. Hahn, and F. Bechstedt: Efficient $\mathcal{O}(N^2)$ method to solve the Bethe-Salpeter equation. In: *Phys. Rev. B* **67** (8), p. 085307 (Feb. 2003), DOI: 10.1103/PhysRevB.67.085307.

[64] W. Brauer: *Einführung in die Elektronentheorie der Metalle*. Akademische Verlagsgesellschaft, Leipzig, 1972.

[65] S. Glutsch: *Excitons in Low-Dimensional Semiconductors: Theory, Numerical Methods, Applications*. Springer, 2004.

[66] Y. Li, X. Luo, and H. Kröger: Bound states and critical behavior of the Yukawa potential. In: *Sci. China Ser. G* **49** (1), pp. 60–71 (2006), DOI: 10.1007/s11433-004-0020-5.

[67] K. Hannewald: Theoretische Untersuchungen zu Paarbildungs- und Kondensationseffekten in optisch angeregten Halbleitern. Diploma thesis, Jena: Friedrich-Schiller-Universität, 1997.

[68] A.-B. Chen and A. Sher: *Semiconductor Alloys*. Plenum, New York, 1995.

[69] J. M. Sanchez, F. Ducastelle, and D. Gratias: Generalized cluster description of multicomponent systems. In: *Physica A* **128** (1-2), pp. 334–350 (1984), DOI: 10.1016/0378-4371(84)90096-7.

[70] A. Zunger: *First-principles statistical mechanics of semiconductor alloys and intermetallic compounds*. Plenum Press, New York 1994, p. 361.

[71] A. Sher, M. van Schilfgaarde, A.-B. Chen, and W. Chen: Quasichemical approximation in binary alloys. In: *Phys. Rev. B* **36** (8), pp. 4279–4295 (Sept. 1987), DOI: 10.1103/PhysRevB.36.4279.

[72] J. W. D. Connolly and A. R. Williams: Density-functional theory applied to phase transformations in transition-metal alloys. In: *Phys. Rev. B* **27** (8), pp. 5169–5172 (Apr. 1983), DOI: 10.1103/PhysRevB.27.5169.

[73] L. K. Teles, J. Furthmüller, L. M. R. Scolfaro, J. R. Leite, and F. Bechstedt: First-principles calculations of the thermodynamic and structural properties of strained $In_xGa_{1-x}N$ and $Al_xGa_{1-x}N$ alloys. In: *Phys. Rev. B* **62** (4), pp. 2475–2485 (July 2000), DOI: 10.1103/PhysRevB.62.2475.

[74] C. Caetano, L. K. Teles, M. Marques, A. Dal Pino, and L. G. Ferreira: Phase stability, chemical bonds, and gap bowing of $In_xGa_{1-x}N$ alloys: Comparison between cubic and wurtzite structures. In: *Phys. Rev. B* **74** (4), p. 045215 (July 2006), DOI: 10.1103/PhysRevB.74.045215.

[75] A. Zunger, S.-H. Wei, L. G. Ferreira, and J. E. Bernard: Special quasirandom structures. In: *Phys. Rev. Lett.* **65** (3), pp. 353–356 (July 1990), DOI: 10.1103/PhysRevLett.65.353.

[76] A. Schleife, M. Eisenacher, C. Rödl, F. Fuchs, J. Furthmüller, and F. Bechstedt: Ab initio description of heterostructural alloys: Thermodynamic and structural properties of $Mg_xZn_{1-x}O$ and $Cd_xZn_{1-x}O$. In: *Phys. Rev. B* **81** (24), p. 245210 (June 2010), DOI: 10.1103/PhysRevB.81.245210.

[77] G. Kresse and J. Furthmüller: Efficient iterative schemes for ab initio total-energy calculations using a plane-wave basis set. In: *Phys. Rev. B* **54** (16), pp. 11169–11186 (Oct. 1996), DOI: 10.1103/PhysRevB.54.11169.

[78] G. Kresse and J. Furthmüller: Efficiency of ab-initio total energy calculations for metals and semiconductors using a plane-wave basis set. In: *Comp. Mater. Sci.* **6** (1), pp. 15–50 (1996), DOI: 10.1016/0927-0256(96)00008-0.

[79] P. E. Blöchl: Projector augmented-wave method. In: *Phys. Rev. B* **50** (24), pp. 17953–17979 (Dec. 1994), DOI: 10.1103/PhysRevB.50.17953.

[80] G. Kresse and D. Joubert: From ultrasoft pseudopotentials to the projector augmented-wave method. In: *Phys. Rev. B* **59** (3), pp. 1758–1775 (Jan. 1999), DOI: 10.1103/PhysRevB.59.1758.

[81] M. Shishkin and G. Kresse: Self-consistent GW calculations for semiconductors and insulators. In: *Phys. Rev. B* **75** (23), p. 235102 (June 2007), DOI: 10.1103/PhysRevB.75.235102.

[82] M. Shishkin, M. Marsman, and G. Kresse: Accurate Quasiparticle Spectra from Self-Consistent GW Calculations with Vertex Corrections. In: *Phys. Rev. Lett.* **99** (24), p. 246403 (Dec. 2007), DOI: 10.1103/PhysRevLett.99.246403.

[83] M. Gajdoš, K. Hummer, G. Kresse, J. Furthmüller, and F. Bechstedt: Linear optical properties in the projector-augmented wave methodology. In: *Phys. Rev. B* **73** (4), p. 045112 (Jan. 2006), DOI: 10.1103/PhysRevB.73.045112.

[84] A. Schleife, C. Rödl, F. Fuchs, J. Furthmüller, F. Bechstedt, P. H. Jefferson, T. D. Veal, C. F. McConville, L. F. J. Piper, A. DeMasi, K. E. Smith, H. Lösch, R. Goldhahn, C. Cobet, J. Zúñiga-Pérez, and V. Muñoz-Sanjosé: *Ab-initio* Studies of Electronic and Spectroscopic Properties of MgO, ZnO, and CdO. In: *J. Korean Phys. Soc.* **53** (5), pp. 2811–2815 (Nov. 2008), DOI: 10.3938/jkps.53.2811.

[85] C. Rödl, F. Fuchs, J. Furthmüller, and F. Bechstedt: Ab initio theory of excitons and optical properties for spin-polarized systems: Application to antiferromagnetic MnO. In: *Phys. Rev. B* **77** (18), p. 184408 (May 2008), DOI: 10.1103/PhysRevB.77.184408.

[86] P. H. Hahn, K. Seino, W. G. Schmidt, J. Furthmüller, and F. Bechstedt: Quasiparticle and excitonic effects in the optical spectra of diamond, SiC, Si, GaP, GaAs, InP, and AlN. In: *Phys. Status Solidi B* **242** (13), pp. 2720–2728 (2005), DOI: 10.1002/pssb.200541128.

[87] C. Rödl, F. Fuchs, J. Furthmüller, and F. Bechstedt: Quasiparticle band structures of the antiferromagnetic transition-metal oxides MnO, FeO, CoO, and NiO. In: *Phys. Rev. B* **79** (23), p. 235114 (June 2009), DOI: 10.1103/PhysRevB.79.235114.

[88] S.-H. Wei and A. Zunger: Role of metal d states in II-VI semiconductors. In: *Phys. Rev. B* **37** (15), pp. 8958–8981 (May 1988), DOI: 10.1103/PhysRevB.37.8958.

[89] A. Schleife, C. Rödl, F. Fuchs, J. Furthmüller, and F. Bechstedt: Optical and energy-loss spectra of MgO, ZnO, and CdO from ab initio many-body calculations. In: *Phys. Rev. B* **80** (3), p. 035112 (July 2009), DOI: 10.1103/PhysRevB.80.035112.

[90] A. Schleife, C. Rödl, F. Fuchs, J. Furthmüller, and F. Bechstedt: Strain influence on valence-band ordering and excitons in ZnO: An ab initio study. In: *Appl. Phys. Lett.* **91** (24), p. 241915 (2007), DOI: 10.1063/1.2825277.

[91] A. Schleife, F. Fuchs, C. Rödl, J. Furthmüller, and F. Bechstedt: Band-structure and optical-transition parameters of wurtzite MgO, ZnO, and CdO from quasiparticle calculations. In: *Phys. Status Solidi B* **246** (9), pp. 2150–2153 (2009), DOI: 10.1002/pssb.200945204.

[92] F. Bechstedt, R. D. Sole, G. Cappellini, and L. Reining: An efficient method for calculating quasiparticle energies in semiconductors. In: *Solid State Commun.* **84** (7), pp. 765–770 (1992), DOI: 10.1016/0038-1098(92)90476-P.

[93] G. Cappellini, R. Del Sole, L. Reining, and F. Bechstedt: Model dielectric function for semiconductors. In: *Phys. Rev. B* **47** (15), pp. 9892–9895 (Apr. 1993), DOI: 10.1103/PhysRevB.47.9892.

[94] H. J. Monkhorst and J. D. Pack: Special points for Brillouin-zone integrations. In: *Phys. Rev. B* **13** (12), pp. 5188–5192 (June 1976), DOI: 10.1103/PhysRevB.13.5188.

[95] R. de L. Kronig: On the theory of dispersion of X-rays. In: *J. Opt. Soc. Am.* **12** (6), pp. 547–556 (1926), DOI: 10.1364/JOSA.12.000547.

[96] H. A. Kramers, in: *Nature* **117**, p. 775 (1926), DOI: 10.1038/117774a0.

[97] T. Makino, K. Tamura, C. H. Chia, Y. Segawa, M. Kawasaki, A. Ohtomo, and H. Koinuma: Optical properties of ZnO:Al epilayers: Observation of room-temperature many-body absorption-edge singularity. In: *Phys. Rev. B* **65** (12), p. 121201 (Mar. 2002), DOI: 10.1103/PhysRevB.65.121201.

[98] B. E. Sernelius, K.-F. Berggren, Z.-C. Jin, I. Hamberg, and C. G. Granqvist: Band-gap tailoring of ZnO by means of heavy Al doping. In: *Phys. Rev. B* **37** (17), pp. 10244–10248 (June 1988), DOI: 10.1103/PhysRevB.37.10244.

[99] S. Major, A. Banerjee, and K. Chopra: Highly transparent and conducting indium-doped zinc oxide films by spray pyrolysis. In: *Thin Solid Films* **108** (3), pp. 333–340 (1983), DOI: 10.1016/0040-6090(83)90082-2.

[100] A. Ohtomo and A. Tsukazaki: Pulsed laser deposition of thin films and superlattices based on ZnO. In: *Semicond. Sci. Tech.* **20** (4), S1–S12 (2005), DOI: 10.1088/0268-1242/20/4/001.

[101] S. Sadofev, S. Blumstengel, J. Cui, J. Puls, S. Rogaschewski, P. Schäfer, Y. G. Sadofyev, and F. Henneberger: Growth of high-quality ZnMgO epilayers and ZnO/ZnMgO quantum well structures by radical-source molecular-beam epitaxy on sapphire. In: *Appl. Phys. Lett.* **87** (9), p. 091903 (2005), DOI: 10.1063/1.2034113.

[102] S. Sadofev, S. Blumstengel, J. Cui, J. Puls, S. Rogaschewski, P. Schäfer, and F. Henneberger: Visible band-gap ZnCdO heterostructures grown by molecular beam epitaxy. In: *Appl. Phys. Lett.* **89** (20), p. 201907 (2006), DOI: 10.1063/1.2388250.

[103] S. Shigemori, A. Nakamura, J. Ishihara, T. Aoki, and J. Temmyo: $Zn_{1-x}Cd_xO$ Film Growth Using Remote Plasma-Enhanced Metalorganic Chemical Vapor Deposition. In: *Jpn. J. Appl. Phys.* **43** (8B), pp. L1088–L1090 (2004), DOI: 10.1143/JJAP.43.L1088.

[104] A. Tsukazaki, A. Ohtomo, T. Kita, Y. Ohno, H. Ohno, and M. Kawasaki: Quantum Hall Effect in Polar Oxide Heterostructures. In: *Science* **315** (5817), pp. 1388–1391 (2007), DOI: 10.1126/science.1137430.

[105] C. Klingshirn, M. Grundmann, A. Hoffmann, B. Meyer, and A. Waag: Zinkoxid – ein alter, neuer Halbleiter. In: *Physik Journal* **1**, pp. 33–38 (Jan. 2006).

[106] L. Schmidt-Mende and J. L. MacManus-Driscoll: ZnO – nanostructures, defects, and devices. In: *Mater. Today* **10** (5), pp. 40–48 (2007), DOI: 10.1016/S1369-7021(07)70078-0.

[107] Y. B. Li, Y. Bando, D. Golberg, and Z. W. Liu: Ga-filled single-crystalline MgO nanotube: Wide-temperature range nanothermometer. In: *Appl. Phys. Lett.* **83** (5), pp. 999–1001 (2003), DOI: 10.1063/1.1597422.

[108] Web of Knowledge. URL: http://www.isiknowledge.com.

[109] W. Martienssen and H. Warlimont, eds.: *Springer Handbook of Condensed Matter and Materials Data*. Springer, Berlin 2005.

[110] B. Gil, A. Lusson, V. Sallet, S.-A. Said-Hassani, R. Triboulet, and P. Bigenwald: Strain-Fields Effects and Reversal of the Nature of the Fundamental Valence Band of ZnO Epilayers. In: *Jpn. J. Appl. Phys.* **40** (Part 2, No. 10B), pp. L1089–L1092 (2001), DOI: 10.1143/JJAP.40.L1089.

[111] Y. Dou, R. G. Egdell, D. S. L. Law, N. M. Harrison, and B. G. Searle: An experimental and theoretical investigation of the electronic structure of CdO. In: *J. Phys. Condens. Mat.* **10** (38), pp. 8447–8458 (1998), DOI: 10.1088/0953-8984/10/38/006.

[112] A. R. H. Preston, B. J. Ruck, L. F. J. Piper, A. DeMasi, K. E. Smith, A. Schleife, F. Fuchs, F. Bechstedt, J. Chai, and S. M. Durbin: Band structure of ZnO from resonant x-ray emission spectroscopy. In: *Phys. Rev. B* **78** (15), p. 155114 (Oct. 2008), DOI: 10.1103/PhysRevB.78.155114.

[113] L. F. J. Piper, A. DeMasi, K. E. Smith, A. Schleife, F. Fuchs, F. Bechstedt, J. Zúñiga-Pérez, and V. Muñoz-Sanjosé: Electronic structure of single-crystal rocksalt CdO studied by soft x-ray spectroscopies and ab initio calculations. In: *Phys. Rev. B* **77** (12), p. 125204 (Mar. 2008), DOI: 10.1103/PhysRevB.77.125204.

[114] P. D. C. King, T. D. Veal, A. Schleife, J. Zúñiga-Pérez, B. Martel, P. H. Jefferson, F. Fuchs, V. Muñoz-Sanjosé, F. Bechstedt, and C. F. McConville: Valence-band electronic structure of CdO, ZnO, and MgO from x-ray photoemission spectroscopy and quasi-particle-corrected density-functional theory calculations. In: *Phys. Rev. B* **79** (20), p. 205205 (May 2009), DOI: 10.1103/PhysRevB.79.205205.

[115] M. Kobayashi, T. Ohkochi, G. S. Song, T. Kataoka, Y. Sakamoto, A. Fujimori, Y. Takeda, T. Okane, Y. Saitoh, H. Yamagami, H. Yamahara, H. Saeki, T. Kawai, and H. Tabata: Experimental observation of bulk band dispersions in the oxide semiconductor ZnO using soft x-ray angle-resolved photoemission spectroscopy. In: *Proceedings of the 29th International Conference on the Physics of Semiconductors* (2008).

[116] A. Kotani and S. Shin: Resonant inelastic x-ray scattering spectra for electrons in solids. In: *Rev. Mod. Phys.* **73** (1), pp. 203–246 (Feb. 2001), DOI: 10.1103/RevModPhys.73.203.

[117] R. T. Girard, O. Tjernberg, G. Chiaia, S. Söderholm, U. O. Karlsson, C. Wigren, H. Nylén, and I. Lindau: Electronic structure of ZnO(0001) studied by angle-resolved photoelectron spectroscopy. In: *Surf. Sci.* **373** (2-3), pp. 409–417 (1997), DOI: 10.1016/S0039-6028(96)01181-8.

[118] R. Enderlein and N. J. M. Horing: *Fundamentals of Semiconductor Physics and Devices*. World Scientific Publishing, 1997.

[119] S. L. Chuang and C. S. Chang: k·p method for strained wurtzite semiconductors. In: *Phys. Rev. B* **54** (4), pp. 2491–2504 (July 1996), DOI: 10.1103/PhysRevB.54.2491.

[120] M. Cardona and N. E. Christensen: Spin-orbit splittings in AlN, GaN and InN. In: *Solid State Commun.* **116** (8), pp. 421–425 (2000), DOI: 10.1016/S0038-1098(00)00353-7.

[121] J. P. Desclaux: Relativistic Dirac-Fock expectation values for atoms with Z = 1 to Z = 120. In: *Atom. Data Nucl. Data* **12** (4), pp. 311–406 (1973), DOI: 10.1016/0092-640X(73)90020-X.

[122] J. Wrzesinski and D. Fröhlich: Two-photon and three-photon spectroscopy of ZnO under uniaxial stress. In: *Phys. Rev. B* **56** (20), pp. 13087–13093 (Nov. 1997), DOI: 10.1103/PhysRevB.56.13087.

[123] A. Mang, K. Reimann, and S. Rübenacke: Band gaps, crystal-field splitting, spin-orbit coupling, and exciton binding energies in ZnO under hydrostatic pressure. In: *Solid State Commun.* **94** (4), pp. 251–254 (1995), DOI: 10.1016/0038-1098(95)00054-2.

[124] B. Gil: Oscillator strengths of A, B, and C excitons in ZnO films. In: *Phys. Rev. B* **64** (20), p. 201310 (Oct. 2001), DOI: 10.1103/PhysRevB.64.201310.

[125] W. R. L. Lambrecht, A. V. Rodina, S. Limpijumnong, B. Segall, and B. K. Meyer: Valence-band ordering and magneto-optic exciton fine structure in ZnO. In: *Phys. Rev. B* **65** (7), p. 075207 (Jan. 2002), DOI: 10.1103/PhysRevB.65.075207.

[126] D. G. Thomas: The exciton spectrum of zinc oxide. In: *J. Phys. Chem. Solids* **15** (1-2), pp. 86–96 (1960), DOI: 10.1016/0022-3697(60)90104-9.

[127] A. V. Rodina, M. Strassburg, M. Dworzak, U. Haboeck, A. Hoffmann, A. Zeuner, H. R. Alves, D. M. Hofmann, and B. K. Meyer: Magneto-optical properties of bound excitons in ZnO. In: *Phys. Rev. B* **69** (12), p. 125206 (Mar. 2004), DOI: 10.1103/PhysRevB.69.125206.

[128] D. C. Reynolds, D. C. Look, B. Jogai, C. W. Litton, G. Cantwell, and W. C. Harsch: Valence-band ordering in ZnO. In: *Phys. Rev. B* **60** (4), pp. 2340–2344 (July 1999), DOI: 10.1103/PhysRevB.60.2340.

[129] Y. Z. Zhu, G. D. Chen, H. Ye, A. Walsh, C. Y. Moon, and S.-H. Wei: Electronic structure and phase stability of MgO, ZnO, CdO, and related ternary alloys. In: *Phys. Rev. B* **77** (24), p. 245209 (June 2008), DOI: 10.1103/PhysRevB.77.245209.

[130] M. Oshikiri, Y. Imanaka, F. Aryasetiawan, and G. Kido: Comparison of the electron effective mass of the n-type ZnO in the wurtzite structure measured by cyclotron resonance and calculated from first principle theory. In: *Physica B* **298** (1-4), pp. 472–476 (2001), DOI: 10.1016/S0921-4526(01)00365-9.

[131] H. Landolt and R. Börnstein: *Numerical Data and Functional Relationships in Science and Technology*. Ed. by K.-H. Hellwege, O. Madelung, M. Schulz, and H. Weiss, vol. III, 17b, Springer-Verlag, Berlin, 1982.

[132] Y.-N. Xu and W. Y. Ching: Self-consistent band structures, charge distributions, and optical-absorption spectra in MgO, α-Al_2O_3, and $MgAl_2O_4$. In: *Phys. Rev. B* **43** (5), pp. 4461–4472 (Feb. 1991), DOI: 10.1103/PhysRevB.43.4461.

[133] N. Ueda, H. Maeda, H. Hosono, and H. Kawazoe: Band-gap widening of CdO thin films. In: *J. Appl. Phys* **84** (11), pp. 6174–6177 (1998), DOI: 10.1063/1.368933.

[134] K. ichi Kawamura, K. Maekawa, H. Yanagi, M. Hirano, and H. Hosono: Observation of carrier dynamics in CdO thin films by excitation with femtosecond laser pulse. In: *Thin Solid Films* **445** (2), pp. 182–185 (2003), Proceedings of the 3rd International Symposium on Transparent Oxide Thin films for Electronics and Optics, DOI: 10.1016/S0040-6090(03)01172-6.

[135] P. H. Jefferson, S. A. Hatfield, T. D. Veal, P. D. C. King, C. F. McConville, J. Zúñiga-Pérez, and V. Muñoz-Sanjosé: Bandgap and effective mass of epitaxial cadmium oxide. In: *Appl. Phys. Lett.* **92** (2), p. 022101 (2008), DOI: 10.1063/1.2833269.

[136] W. Mönch: *Semiconductor Surfaces and Interfaces*. Springer, Berlin 2001.

[137] F. Flores and C. Tejedor: Energy barriers and interface states at heterojunctions. In: *J. Phys. C Solid State* **12** (4), pp. 731–749 (1979), DOI: 10.1088/0022-3719/12/4/018.

[138] J. Tersoff: Theory of semiconductor heterojunctions: The role of quantum dipoles. In: *Phys. Rev. B* **30** (8), pp. 4874–4877 (Oct. 1984), DOI: 10.1103/PhysRevB.30.4874.

[139] C. G. Van de Walle and J. Neugebauer: Universal alignment of hydrogen levels in semiconductors, insulators and solutions. In: *Nature* **423**, pp. 626–628 (2003), DOI: 10.1038/nature01665.

[140] M. Eisenacher, F. Fuchs, and C. Rödl, private communication, 2008.

[141] I. Mahboob, T. D. Veal, C. F. McConville, H. Lu, and W. J. Schaff: Intrinsic Electron Accumulation at Clean InN Surfaces. In: *Phys. Rev. Lett.* **92** (3), p. 036804 (Jan. 2004), DOI: 10.1103/PhysRevLett.92.036804.

[142] I. Mahboob, T. D. Veal, L. F. J. Piper, C. F. McConville, H. Lu, W. J. Schaff, J. Furthmüller, and F. Bechstedt: Origin of electron accumulation at wurtzite InN surfaces. In: *Phys. Rev. B* **69** (20), p. 201307 (May 2004), DOI: 10.1103/PhysRevB.69.201307.

[143] C. Tejedor and F. Flores: A simple approach to heterojunctions. In: *J. Phys. C Solid State* **11** (1), pp. L19–L23 (1977), DOI: 10.1088/0022-3719/11/1/005.

[144] M. Cardona and N. E. Christensen: Acoustic deformation potentials and heterostructure band offsets in semiconductors. In: *Phys. Rev. B* **35** (12), pp. 6182–6194 (Apr. 1987), DOI: 10.1103/PhysRevB.35.6182.

[145] A. Schleife, F. Fuchs, C. Rödl, J. Furthmüller, and F. Bechstedt: Branch-point energies and band discontinuities of III-nitrides and III-/II-oxides from quasiparticle band-structure calculations. In: *Appl. Phys. Lett.* **94** (1), p. 012104 (2009), DOI: 10.1063/1.3059569.

[146] T. D. Veal, P. D. C. King, S. A. Hatfield, L. R. Bailey, C. F. McConville, B. Martel, J. C. Moreno, E. Frayssinet, F. Semond, and J. Zúñiga-Pérez: Valence band offset of the ZnO/AlN heterojunction determined by x-ray photoemission spectroscopy. In: *Appl. Phys. Lett.* **93** (20), p. 202108 (2008), DOI: 10.1063/1.3032911.

[147] L. Piper, P. Jefferson, T. Veal, C. McConville, J. Zúñiga-Pérez, and V. Muñoz-Sanjosé: X-ray photoemission studies of the electronic structure of single-crystalline CdO(100). In: *Superlattices Microst.* **42** (1-6), pp. 197–200 (2007), E-MRS 2006 Symposium K: ZnO and Related Materials, 2006 Spring Meeting of the European Materials Research Society, DOI: 10.1016/j.spmi.2007.04.029.

[148] P. D. C. King, T. D. Veal, P. H. Jefferson, J. Zúñiga-Pérez, V. Muñoz-Sanjosé, and C. F. McConville: Unification of the electrical behavior of defects, impurities, and surface states in semiconductors: Virtual gap states in CdO. In: *Phys. Rev. B* **79** (3), p. 035203 (Jan. 2009), DOI: 10.1103/PhysRevB.79.035203.

[149] L. F. J. Piper, L. Colakerol, P. D. C. King, A. Schleife, J. Zúñiga-Pérez, P.-A. Glans, T. Learmonth, A. Federov, T. D. Veal, F. Fuchs, V. Muñoz-Sanjosé, F. Bechstedt, C. F. McConville, and K. E. Smith: Observation of quantized subband states and evidence for surface electron accumulation in CdO from angle-resolved photoemission spectroscopy. In: *Phys. Rev. B* **78** (16), p. 165127 (Oct. 2008), DOI: 10.1103/PhysRevB.78.165127.

[150] M. L. Bortz, R. H. French, D. J. Jones, R. V. Kasowski, and F. S. Ohuchi: Temperature dependence of the electronic structure of oxides: MgO, $MgAl_2O_4$ and Al_2O_3. In: *Phys. Scr.* **41** (4), pp. 537–541 (1990), DOI: 10.1088/0031-8949/41/4/036.

[151] M. W. Williams and E. T. Arakawa: Optical Properties of Single-Crystal Magnesium Oxide. In: *J. Appl. Phys.* **38** (13), pp. 5272–5276 (1967), DOI: 10.1063/1.1709313.

[152] D. M. Roessler and W. C. Walker: Electronic Spectrum and Ultraviolet Optical Properties of Crystalline MgO. In: *Phys. Rev.* **159** (3), pp. 733–738 (July 1967), DOI: 10.1103/PhysRev.159.733.

[153] N.-P. Wang, M. Rohlfing, P. Krüger, and J. Pollmann: Electronic excitations of CO adsorbed on MgO(001). In: *Appl. Phys. A* **78** (2), pp. 213–221 (Jan. 2004), DOI: 10.1007/s00339-003-2305-3.

[154] M. Rakel, C. Cobet, N. Esser, P. Gori, O. Pulci, A. Seitsonen, A. Cricenti, N. Nickel, and W. Richter: *Epioptics-9*. World Scientific, 2008, chap. Electronic and Optical Properties of ZnO Between 3 and 32 eV.

[155] R. Klucker, H. Nelkowski, Y. S. Park, M. Skibowski, and T. S. Wagner: Optical anisotropy of ZnO in the ultraviolet region. In: *Phys. Status Solidi B* **45** (1), pp. 265–272 (1971), DOI: 10.1002/pssb.2220450130.

[156] R. Laskowski and N. E. Christensen: Ab initio calculation of excitons in ZnO. In: *Phys. Rev. B* **73** (4), p. 045201 (Jan. 2006), DOI: 10.1103/PhysRevB.73.045201.

[157] J. L. Freeouf: Far-Ultraviolet Reflectance of II-VI Compounds and Correlation with the Penn—Phillips Gap. In: *Phys. Rev. B* **7** (8), pp. 3810–3830 (Apr. 1973), DOI: 10.1103/PhysRevB.7.3810.

[158] R. J. Elliott: Intensity of Optical Absorption by Excitons. In: *Phys. Rev.* **108** (6), pp. 1384–1389 (Dec. 1957), DOI: 10.1103/PhysRev.108.1384.

[159] F. Bechstedt, K. Seino, P. H. Hahn, and W. G. Schmidt: Quasiparticle bands and optical spectra of highly ionic crystals: AlN and NaCl. In: *Phys. Rev. B* **72** (24), p. 245114 (Dec. 2005), DOI: 10.1103/PhysRevB.72.245114.

[160] R. C. Whited, C. J. Flaten, and W. C. Walker: Exciton thermoreflectance of MgO and CaO. In: *Solid State Commun.* **13** (11), pp. 1903–1905 (1973), DOI: 10.1016/0038-1098(73)90754-0.

[161] G. F. Koster, J. O. Dimmock, R. G. Wheeler, and H. Statz: *Properties of the thirty-two point groups*. M.I.T. Press, Cambridge, Massachusetts, 1963.

[162] S. Kohiki, M. Arai, H. Yoshikawa, and S. Fukushima: Energy Loss Structure of X-ray Photoelectron Spectra of MgO and α-Al_2O_3. In: *J. Phys. Chem. B* **103** (25), pp. 5296–5299 (1999), DOI: 10.1021/jp9901996.

[163] M. R. Wagner, M. Cobet, G. Callsen, and A. Hofmann, private communication, 2010.

[164] C. Bundesmann, A. Rahm, M. Lorenz, M. Grundmann, and M. Schubert: Infrared optical properties of $Mg_xZn_{1-x}O$ thin films ($0 \leq x \leq 1$): Long-wavelength optical phonons and dielectric constants. In: *J. Appl. Phys.* **99** (11), p. 113504 (2006), DOI: 10.1063/1.2200447.

[165] Z. Liu, Z. Mei, T. Zhang, Y. Liu, Y. Guo, X. Du, A. Hallen, J. Zhu, and A. Kuznetsov: Solar-blind 4.55 eV band gap $Mg_{0.55}Zn_{0.45}O$ components fabricated using quasi-homo buffers. In: *J. Cryst. Growth* **311** (18), pp. 4356–4359 (2009), DOI: 10.1016/j.jcrysgro.2009.07.030.

[166] X. Du, Z. Mei, Z. Liu, Y. Guo, T. Zhang, Y. Hou, Z. Zhang, Q. Xue, and A. Y. Kuznetsov: Controlled Growth of High-Quality ZnO-Based Films and Fabrication of Visible-Blind and Solar-Blind Ultra-Violet Detectors. In: *Adv. Mater.* **21** (45), pp. 4625–4630 (2009), DOI: 10.1002/adma.200901108.

[167] J. Furthmüller, P. H. Hahn, F. Fuchs, and F. Bechstedt: Band structures and optical spectra of InN polymorphs: Influence of quasiparticle and excitonic effects. In: *Phys. Rev. B* **72** (20), p. 205106 (Nov. 2005), DOI: 10.1103/PhysRevB.72.205106.

[168] P. Rinke, M. Winkelnkemper, A. Qteish, D. Bimberg, J. Neugebauer, and M. Scheffler: Consistent set of band parameters for the group-III nitrides AlN, GaN, and InN. In: *Phys. Rev. B* **77** (7), p. 075202 (Feb. 2008), DOI: 10.1103/PhysRevB.77.075202.

[169] R. Schmidt, B. Rheinländer, M. Schubert, D. Spemann, T. Butz, J. Lenzner, E. M. Kaidashev, M. Lorenz, A. Rahm, H. C. Semmelhack, and M. Grundmann: Dielectric functions (1 to 5 eV) of wurtzite $Mg_xZn_{1-x}O$ ($x \leq 0.29$) thin films. In: *Appl. Phys. Lett.* **82** (14), pp. 2260–2262 (2003), DOI: 10.1063/1.1565185.

[170] M. Sanati, G. L. W. Hart, and A. Zunger: Ordering tendencies in octahedral MgO-ZnO alloys. In: *Phys. Rev. B* **68** (15), p. 155210 (Oct. 2003), DOI: 10.1103/PhysRevB.68.155210.

[171] S.-H. Jeong, B.-S. Kim, and B.-T. Lee: Photoluminescence dependence of ZnO films grown on Si(100) by radio-frequency magnetron sputtering on the growth ambient. In: *Appl. Phys. Lett.* **82** (16), pp. 2625–2627 (2003), DOI: 10.1063/1.1568543.

[172] J. Chen, W. Z. Shen, N. B. Chen, D. J. Qiu, and H. Z. Wu: The study of composition non-uniformity in ternary $Mg_xZn_{1-x}O$ thin films. In: *J. Phys. Condens. Mat.* **15** (30), p. L475 (2003), DOI: 10.1088/0953-8984/15/30/102.

[173] F. Bertram, S. Giemsch, D. Forster, J. Christen, R. Kling, C. Kirchner, and A. Waag: Direct imaging of phase separation in ZnCdO layers. In: *Appl. Phys. Lett.* **88** (6), p. 061915 (2006), DOI: 10.1063/1.2172146.

[174] T. Makino, Y. Segawa, M. Kawasaki, A. Ohtomo, R. Shiroki, K. Tamura, T. Yasuda, and H. Koinuma: Band gap engineering based on $Mg_xZn_{1-x}O$ and $Cd_yZn_{1-y}O$ ternary alloy films. In: *Appl. Phys. Lett.* **78** (9), pp. 1237–1239 (2001), DOI: 10.1063/1.1350632.

[175] T. Minemoto, T. Negami, S. Nishiwaki, H. Takakura, and Y. Hamakawa: Preparation of $Zn_{1-x}Mg_xO$ films by radio frequency magnetron sputtering. In: *Thin Solid Films* **372** (1-2), pp. 173–176 (2000), DOI: 10.1016/S0040-6090(00)01009-9.

[176] Z. Vashaei, T. Minegishi, H. Suzuki, T. Hanada, M. W. Cho, T. Yao, and A. Setiawan: Structural variation of cubic and hexagonal $Mg_xZn_{1-x}O$ layers grown on MgO(111)/c-sapphire. In: *J. Appl. Phys.* **98** (5), p. 054911 (2005), DOI: 10.1063/1.2039273.

[177] X. F. Fan, H. D. Sun, Z. X. Shen, J.-L. Kuo, and Y. M. Lu: A first-principle analysis on the phase stabilities, chemical bonds and band gaps of wurtzite structure $A_xZn_{1-x}O$ alloys (A = Ca, Cd, Mg). In: *J. Phys. Condens. Mat.* **20** (23), p. 235221 (2008), DOI: 10.1088/0953-8984/20/23/235221.

[178] J. Ishihara, A. Nakamura, S. Shigemori, T. Aoki, and J. Temmyo: $Zn_{1-x}Cd_xO$ systems with visible band gaps. In: *Appl. Phys. Lett.* **89** (9), p. 091914 (2006), DOI: 10.1063/1.2345232.

[179] A. Ohtomo, M. Kawasaki, T. Koida, K. Masubuchi, H. Koinuma, Y. Sakurai, Y. Yoshida, T. Yasuda, and Y. Segawa: $Mg_xZn_{1-x}O$ as a II–VI widegap semiconductor alloy. In: *Appl. Phys. Lett.* **72** (19), pp. 2466–2468 (1998), DOI: 10.1063/1.121384.

[180] W. R. L. Lambrecht, S. Limpijumnong, and B. Segall: Theoretical studies of ZnO and related $Mg_xZn_{1-x}O$ alloy band structures. In: *MRS Internet J. Nitride Semicond. Res.* **4S1**, G6.8 (1999).

[181] D. Fritsch, H. Schmidt, and M. Grundmann: Pseudopotential band structures of rocksalt MgO, ZnO, and $Mg_{1-x}Zn_xO$. In: *Appl. Phys. Lett.* **88** (13), p. 134104 (2006), DOI: 10.1063/1.2188382.

[182] R. K. Thöt, T. Sander, P. J. Klar, and B. Meyer: Photocurrent measurements of $Mg_xZn_{1-x}O$ epitaxial layers of different x. In: *MRS Proceedings* **1201**, 1201–H10–33 (2009), DOI: 10.1557/PROC-1201-H10-33.

[183] L. A. Kappers, R. L. Kroes, and E. B. Hensley: $F+$ and F' Centers in Magnesium Oxide. In: *Phys. Rev. B* **1** (10), pp. 4151–4157 (May 1970), DOI: 10.1103/PhysRevB.1.4151.

[184] Y. Chen, V. M. Orera, R. Gonzalez, R. T. Williams, G. P. Williams, G. H. Rosenblatt, and G. J. Pogatshnik: Effect of substitutional hydride ions on the charge states of oxygen vacancies in thermochemically reduced CaO and MgO. In: *Phys. Rev. B* **42** (2), pp. 1410–1416 (July 1990), DOI: 10.1103/PhysRevB.42.1410.

[185] J. E. Wertz, P. Auzins, R. A. Weeks, and R. H. Silsbee: Electron Spin Resonance of F Centers in Magnesium Oxide; Confirmation of the Spin of Magnesium-25. In: *Phys. Rev.* **107** (6), pp. 1535–1537 (Sept. 1957), DOI: 10.1103/PhysRev.107.1535.

[186] Y. Chen, R. T. Williams, and W. A. Sibley: Defect Cluster Centers in MgO. In: *Phys. Rev.* **182** (3), pp. 960–964 (June 1969), DOI: 10.1103/PhysRev.182.960.

[187] M. Hedström, A. Schindlmayr, G. Schwarz, and M. Scheffler: Quasiparticle Corrections to the Electronic Properties of Anion Vacancies at GaAs(110) and InP(110). In: *Phys. Rev. Lett.* **97** (22), p. 226401 (Nov. 2006), DOI: 10.1103/PhysRevLett.97.226401.

[188] P. Rinke, A. Janotti, A. Schleife, E. Kioupakis, C. Rödl, F. Bechstedt, C. G. Van de Walle, and M. Scheffler, In preparation,

[189] Y. Qimin, P. Rinke, M. Scheffler, and C. G. Van de Walle, In preparation,

[190] C. Freysoldt, J. Neugebauer, and C. G. Van de Walle: Fully Ab Initio Finite-Size Corrections for Charged-Defect Supercell Calculations. In: *Phys. Rev. Lett.* **102** (1), p. 016402 (Jan. 2009), DOI: 10.1103/PhysRevLett.102.016402.

[191] U. Özgür, Y. I. Alivov, C. Liu, A. Teke, M. A. Reshchikov, S. Doğan, V. Avrutin, S.-J. Cho, and H. Morkoç: A comprehensive review of ZnO materials and devices. In: *J. Appl. Phys.* **98** (4), p. 041301 (2005), DOI: 10.1063/1.1992666.

[192] C. Klingshirn, R. Hauschild, J. Fallert, and H. Kalt: Room-temperature stimulated emission of ZnO: Alternatives to excitonic lasing. In: *Phys. Rev. B* **75** (11), p. 115203 (Mar. 2007), DOI: 10.1103/PhysRevB.75.115203.

[193] E. Burstein: Anomalous Optical Absorption Limit in InSb. In: *Phys. Rev.* **93** (3), pp. 632–633 (Feb. 1954), DOI: 10.1103/PhysRev.93.632.

[194] K. Hannewald, S. Glutsch, and F. Bechstedt: Quantum-kinetic study of femtosecond pump-and-probe spectra of bulk GaAs. In: *Phys. Rev. B* **61** (16), pp. 10792–10802 (Apr. 2000), DOI: 10.1103/PhysRevB.61.10792.

[195] G. D. Mahan: Excitons in Degenerate Semiconductors. In: *Phys. Rev.* **153** (3), pp. 882–889 (Jan. 1967), DOI: 10.1103/PhysRev.153.882.

[196] K. F. Berggren and B. E. Sernelius: Band-gap narrowing in heavily doped many-valley semiconductors. In: *Phys. Rev. B* **24** (4), pp. 1971–1986 (Aug. 1981), DOI: 10.1103/PhysRevB.24.1971.

[197] J. Wu, W. Walukiewicz, W. Shan, K. M. Yu, J. W. Ager, E. E. Haller, H. Lu, and W. J. Schaff: Effects of the narrow band gap on the properties of InN. In: *Phys. Rev. B* **66** (20), p. 201403 (Nov. 2002), DOI: 10.1103/PhysRevB.66.201403.

[198] M. Eisenacher: Eigenschaften von $Mg_xZn_{1-x}O$: Parameterfreie Untersuchungen. Diploma thesis, Jena: Friedrich-Schiller-Universität, 2008.

[199] F. D. Murnaghan: The Compressibility of Media under Extreme Pressures. In: *P. Nat. Acad. Sci. USA* **30** (9), pp. 244–247 (1944), URL: http://www.pnas.org/content/30/9/244.short.

Acknowledgments

I wish to thank everyone whose enduring support during my time as PhD student has made this work possible.

Most importantly, I want to thank Prof. Dr. Friedhelm Bechstedt. His ideas and his scientific inspiration motivated me throughout my studies. Due to his efforts, my financial situation was always stable and I never had to worry about the computational infrastructure. Lively discussions with him guided me on the scientific journeys that led to this work.

I would also like to express my gratitude towards my former colleagues from the Institut für Festkörpertheorie und -optik in Jena for the stimulating debates and their help on which I could always rely.

I am very grateful to Prof. Chris G. Van de Walle for his cordial hospitality. He enabled my seminal stay at the University of California, Santa Barbara, where I experienced the motivating, productive, and friendly atmosphere which is omnipresent in his group.

Numerous collaborators, both experimentalists and theoreticians, share with me a fascination in research related to topics discussed in this work. In this context I want to thank Dr. Christoph Cobet, Dr. Munise Cobet, Prof. Dr. Rüdiger Goldhahn, Dr. Emmanouil Kioupakis, Dr. Anderson Janotti, Dr. Louis F. J. Piper, Dr. Anna Rodina, Dr. Joel Varley, Dr. Tim Veal, and Dr. Markus R. Wagner for stimulating discussions, good suggestions, and sharing data.

Working on scientific problems with Dr. Claudia Rödl and Dr. Frank Fuchs made me more experienced and patient. Our many conversations, work-related and personal, and knowing I could always rely on their help and opinions has made me value them as both friends and colleagues. In addition, I want to thank Dr. Patrick Rinke and Thiago Pantaleão Melo who I met in Santa Barbara and whose friendship led to fruitful and ongoing developments in my scientific and private life.

I want to thank my family and friends for their support of my work and their trust in me. In particular, I am thankful to my girlfriend Yvonne Gower. Her love and her presence in my life are the source of the energy and the inspiration that was necessary to accomplish this work.

I am grateful for scholarships from the Carl-Zeiss-Stiftung and the Free State of Thuringia which sponsored and promoted my work.

Die VDM Verlagsservicegesellschaft sucht für wissenschaftliche Verlage abgeschlossene und herausragende

Dissertationen, Habilitationen, Diplomarbeiten, Master Theses, Magisterarbeiten usw.

für die kostenlose Publikation als Fachbuch.

Sie verfügen über eine Arbeit, die hohen inhaltlichen und formalen Ansprüchen genügt, und haben Interesse an einer honorarvergüteten Publikation?

Dann senden Sie bitte erste Informationen über sich und Ihre Arbeit per Email an *info@vdm-vsg.de*.

Sie erhalten kurzfristig unser Feedback!

VDM Verlagsservicegesellschaft mbH
Dudweiler Landstr. 99
D - 66123 Saarbrücken

Telefon +49 681 3720 174
Fax   +49 681 3720 1749

www.vdm-vsg.de

Die VDM Verlagsservicegesellschaft mbH vertritt

Printed by Books on Demand GmbH, Norderstedt / Germany